International Political Economy Series

Series Editor
Timothy M. Shaw
Visiting Professor
University of Massachusetts Boston, USA

Emeritus Professor
University of London, UK

The global political economy is in flux as a series of cumulative crises impacts its organization and governance. The IPE series has tracked its development in both analysis and structure over the last three decades. It has always had a concentration on the global South. Now the South increasingly challenges the North as the centre of development, also reflected in a growing number of submissions and publications on indebted Eurozone economies in Southern Europe. An indispensable resource for scholars and researchers, the series examines a variety of capitalisms and connections by focusing on emerging economies, companies and sectors, debates and policies. It informs diverse policy communities as the established trans-Atlantic North declines and 'the rest', especially the BRICS, rise.

More information about this series at
http://www.springer.com/series/13996

'In a clear and powerful argument, grounded in evidence from a wide range of sources, this book will persuade you that only eco-socialism can solve the climate crisis and ensure our survival.'

—Jacklyn Cock, *University of the Witwatersrand, South Africa*

Peadar Kirby • Tadhg O'Mahony

The Political Economy of the Low-Carbon Transition

Pathways Beyond Techno-Optimism

palgrave
macmillan

Peadar Kirby
Department of Politics and Public Administration
University of Limerick
Limerick, Ireland

Tadhg O'Mahony
Finland Futures Research Centre
Turku School of Economics
University of Turku
Tampere, Finland

International Political Economy Series
ISBN 978-3-319-62553-9 ISBN 978-3-319-62554-6 (eBook)
DOI 10.1007/978-3-319-62554-6

Library of Congress Control Number: 2017947768

Printed on acid-free paper

This Palgrave Macmillan imprint is published by Springer Nature
The registered company is Springer International Publishing AG
The registered company address is: Gewerbestrasse 11, 6330 Cham, Switzerland

Peadar:

To the friend I have had longest in my life, Eamonn O'Dwyer, his partner Tessa and his family, Jess, Sam and Alannah:

Eamonn was diagnosed with pancreatic cancer as I began writing this book and was journeying with great dignity and love towards his death, surrounded by his family, as I finished writing. Eamonn is a soul friend whom I shall miss greatly after more than 50 years of deeply nourishing encounters. With great love, I dedicate this book to Eamonn, Tessa, Jess, Sam and Alannah: you will always be with us Eamonn.

Tadhg:

To my Uncle John O'Mahony, his wife Peigí and family Máire, Aoife, Mossy, Johnny, Diarmuid, David and the late Timmy. To my Aunty Kathleen Meagher, her husband Ned and daughter Aoife:

Uncle John passed away in June 2016 and Aunty Kathleen in November. This book is about a journey. I have many memories of journeys that involved them in some way. From the Devil's Bit to Watergrasshill. I dedicate this book to your memory.

Foreword

Unchecked climate change is a looming existential threat. Yet lifting the bonnet on the mitigation agenda proselytised by many climate elites reveals no meaningful nor timely action to curb emissions in line with our Paris commitments. Instead, salvation is to be found in a plethora of glossy reports promoting green-growth, higher efficiency, utopian technology and the financialisation of all we hold dear.

The latest report from the Intergovernmental Panel on Climate Change combined with the obligations enshrined in the Paris Agreement of 2015 has reshaped the climate change agenda. While the former establishes carbon budgets as the appropriate scientific foundation for mitigation policy, the latter requires the international community 'to hold the increase in global average temperature to well below 2°C' and to 'pursue efforts to limit the temperature increase to 1.5°C'.

This ambitious and scientifically informed agenda demands rates of mitigation without historical precedent that are unimaginable within contemporary politics and remain far beyond anything yet countenanced across mainstream academia. Even a conservative reading of the Paris commitments requires the wholesale transformation of the global energy system from high carbon fossil fuels to zero carbon alternatives by, if not well before, 2050.

It is to this Herculean challenge that Peadar Kirby and Tadhg O'Mahony's attention is focused in their *The Political Economy of the Low-Carbon Transition: Pathways beyond Techno-Optimism*.

Clear and engaging, Peadar and Tadhg guide the reader through the many facets of climate change, from a contextual characterisation of the problem, its place within the wider sustainability discourse, and from what social and economic structures judicious solutions may arise.

Technical and scientific issues are adequately covered, as are the various critiques. However, this book makes no pretence to substitute for 101 climate science and its value is certainly enhanced if the reader is already familiar with climate science and the concept of carbon budgets. The real strength of the analysis is in situating the technocratic framing of climate change within an explicit and evolving political and social context.

The authors' perspective and preferences are clear—they maintain a critical perspective throughout—ultimately providing an open interpretation of the challenges faced and of the potential responses and solutions. They are evidently unconvinced by the technocratic and market-mechanism responses to climate change, seeing them very much as part of the problem rather than a framework for solutions. Their views here are well constructed and emerge from a clear understanding of the historical timeline that has delivered both contemporary society and its accompanying problems. The transparency of their reasoning makes it an appropriate and valuable read for those concerned about climate change, but who interpret the mitigation landscape through a more conventional lens. In this respect their analysis opens up the prospect for informed debate—from which a richer understanding of the challenges should emerge—even if disagreement still remains.

In constructing their arguments Peadar and Tadhg draw on experiences from international development to shed light on the dynamic interplay between technology, politics, culture, economics and power. In contrast to much of the academic guidance on mitigation, they demonstrate a deep appreciation of political economy and its pivotal role in thwarting or driving any meaningful progress.

Sadly, the growing dominance of abstract and quantitative scenarios generated by ever more complex and black box modelling has increasingly sidelined the thorny issues exposed by an understanding of political economy. Such 'expert-based' and highly technical approaches have effectively closed down debate, providing instead inadequate responses to climate change that do not threaten the dominant socioeconomic paradigm.

Eloquently capturing this process of marginalising plurality, Peadar and Tadhg turn to the wisdom of Pope Francis who writes in his encyclical letter *Laudato Sí* that 'the alliance of technology and economics ends up side-lining anything unrelated to its immediate interests. Consequently the most one can expect is superficial rhetoric, sporadic acts of philanthropy, and perfunctory expressions of concern for the environment, whereas any genuine attempt to introduce change is viewed as a nuisance based on romantic illusions.'

Building on this Peadar and Tadhg discuss groups grappling with how to operationalise the rich world latent in the Pope's 'romantic illusions'. They draw attention to how such approaches offer alternative and often contrasting visions rather than the singularity forthcoming from the dominant modelling approaches (technically referred to as Integrated Assessment Models—IAMs). They also emphasise how the distinction between transition and transformation is much more than semantics: the former captures a programme of incremental *adjustments within* the contemporary paradigm, while the latter embeds change that is fundamentally *challenging to* the paradigm.

Peadar and Tadhg offer the reader an informed tour of the prominent landmarks scattered across the climate change landscape, though their principal contribution is in revealing the often opaque links between them. In this regard the book is appropriate for a wide constituency of readers. The well-informed climate scientist will be enlightened through discussions on power, equity and the thorny issues residing in the social and political sciences. At the same time, those with a good grasp of political shenanigans, power struggles and competitive commerce can witness the tortuous and time-consuming path climate change has had to navigate to become a pivotal global issue—as well as the trials and tribulations that continue to thwart meaningful and timely action. In many respects, *The Political Economy of the Low-Carbon Transition* is an excellent undergraduate text, enriching the understanding of those studying the more technical elements of climate change and providing a useful and in-depth reference for those with an interdisciplinary bent to their studies.

This book is an important contribution on at least two key levels. First it documents how *the alliance of technology and economics [is] side-lining anything unrelated to its immediate interests.* And second, it details how the success of tomorrow's reality can be found deep in the transformations hidden in today's *romantic illusions.*

The future will be radically different from today. We either continue the mitigation masquerade and face the chaotic consequences of rapidly rising temperatures, or we cull the neo-liberal model and begin a radical mitigation agenda based on integrity and equity. The window for deciding on which future to bequeath our children is almost closed, but for today at least, the choice is ours.

Kevin Anderson

Professor of Energy and Climate Change
University of Manchester

Deputy Director of the Tyndall
Centre for Climate Change Research

Contents

Abbreviations

ACCCRN	Asian Cities Climate Change Resilience Network
ADB	Asian Development Bank
AGN	African Group of Negotiators
AILAC	Independent Association of Latin America and the Caribbean
ALBA	Bolivarian Alliance for the Peoples of our America
AME	Asian Modelling Exercise
AOSIS	Alliance of Small Island States
AR4	Fourth Assessment Report of the IPCC
AR5	Fifth Assessment Report of the IPCC
BAU	Business-as-usual
BP	British Petroleum
CAIT	Climate Access Indicators Tool of the World Resources Institute
CAT	Centre for Alternative Technology
CBDR	Common But Differentiated Responsibilities
CCS	Carbon Capture and Storage
CDM	Clean Development Mechanism
CDR	Carbon dioxide removal
CEPAL	UN Economic Commission for Latin America
CEPGs	Climate and environmental policy groups
CFCs	Chlorofluorocarbons
CH4	Methane
CME	Coordinated market economy
CO_2	Carbon dioxide
CoP	Conference of the Parties
CSR	Corporate Social Responsibility
DAI	Dangerous Anthropogenic Interference
DDT	Dichlorodiphenyltrichloroethane

DRC	Democratic Republic of the Congo
EC	European Commission
EEA	European Environment Agency
EESC	European Economic and Social Committee
EF	Ecological Footprint
EKC	Environmental Kuznets Curve
ERI	Environmental Research Institute
ETS	Emissions Trading Schemes
EU	European Union
FAO	Food and Agriculture Organisation
GCEC	Global Commission on the Economy and Climate
GDP	Gross domestic product
GEG	Global energy governance
GEO	Global Environmental Outlook
GHG	Greenhouse gas
GT	Great Transition
ICTA	Institute of Environmental Science and Technology
IEA	International Energy Agency
IMF	International Monetary Fund
INDC	Intended Nationally Determined Contribution
IOPN	International Office for the Protection of Nature
IPAT	$I = P \times A \times T$, where I is the environmental impact, P is population, A is affluence and T is technology
IPCC	Intergovernmental Panel on Climate Change
IPSP	International Panel on Social Progress
IS92	Integrated science scenarios published by the IPCC in 1992
ISEW	Index of Sustainable Economic Welfare
ISI	Import Substitution Industrialisation
ITF	International Transport Forum
IUPN	International Union for the Protection of Nature
KCL	King's College London
LMDC	Like-Minded Developing Countries
LME	Liberal market economy
LSE	London School of Economics
LULUCF	Landuse, Landuse Change and Forestry
MAXWELL	Maximise wellbeing, minimise emissions
MEA	Millennium Ecosystem Assessment
MF	Market Forces
MIT	Massachusetts Institute of Technology
MRFCJ	Mary Robinson Foundation—Climate Justice
N_20	Nitrous Oxide
NASA	National Aeronautics and Space Administration

NEA	Nuclear Energy Agency
NEF	New Economics Foundation
NESTI	National Energy and Environment Strategy for Technological Innovation
NGO	Non-governmental organisation
NIEO	New International Economic Order
NOAA	National Oceanic and Atmospheric Administration
NO_x	Nitrogen oxides
ODI	Overseas Development Institute
OECD	Organisation for Economic Cooperation and Development
OLCA	Latin American Observatory of Environmental Conflicts
OPEC	Organisation of the Petroleum Exporting Countries
PA	Paris Agreement
PB	Planetary boundaries
PP	Partido Popular
PPLD	Parti Pour La Décroissance
ppm	Parts per million
QDI	Quality of Development Index
QUELROs	Quantified emissions limitation and reduction objectives
R&D	Research and Development
RICO	Racketeer Influenced Corrupt Organisations
RSP	Regulatory State Paradigm
SCC	Social Cost of Carbon
SCP	Sustainable Consumption and Production
SD	Sustainable development
SDG	Sustainable Development Goal
SIDS	Small island developing states
SO_2	Sulphur dioxide
SOAS	School of Oriental and African Studies
SRES	Special Report on Emission Scenarios
SSP	Shared Socioeconomic Pathway
TAR	Third Assessment Report of the IPCC
TFC	The final consumption of energy
TIMES	The Integrated MARKAL-EFOM System model generator
TIPNIS	Isiboro Sécure indigenous National Park, Bolivia
TNC	Transnational Corporation
TPES	Total primary energy supply
UK	United Kingdom
UN	United Nations
UNDP	United Nations Development Programme
UNECLAC	United Nations Economic Commission for Latin America and the Caribbean

UNEP	United Nations Environment Programme
UNESCAP	UN Economic and Social Commission for Asia and the Pacific
UNFCCC	United Nations Framework Convention on Climate Change
US	United States
WBGU	German Advisory Council on Global Change
WCED	World Commission on Environment and Development
WMO	World Meteorological Organisation
WWF	World Wildlife Fund
WWS	Wind-Water-Sun

List of Figures

List of Tables

PART I

Climate Change as Problem

CHAPTER 1

Defining the Problem: The Complex Dimensions of the Grave New Threats We Face

Introduction

By the normal practices of international politics, many of the speeches made by world leaders at the Paris climate summit in December 2015 were exceptional. Not only was this the largest meeting of heads of state and government from all over the world that had ever taken place, but it was used to recognise that the growing threat of climate change could, as President Barack Obama put it, 'define the contours of this century more dramatically than any other', resulting in 'submerged countries, abandoned cities, fields that no longer grow, political disruptions that trigger new conflict, and even more floods of desperate peoples seeking the sanctuary of nations not their own'. The summit's host, President François Hollande, put it even more starkly: 'never—truly never—have the stakes of an international meeting been so high. For the future of the planet, and the future of life, are at stake'. UN Secretary General Ban Ki-moon agreed: 'We have never faced such a test. … Paris must mark a turning point. We need the world to know that we are headed to a low-emissions, climate-resilient future, and that there is no going back.'

This is not the first time that humanity faced global catastrophe: the threat of nuclear annihilation loomed over the Cold War world. But the world's leaders were correct in describing the threat now facing us as the greatest test ever since it is caused not by the triggering

P. Kirby, T. O'Mahony, *The Political Economy of the Low-Carbon Transition*, International Political Economy Series,
DOI 10.1007/978-3-319-62554-6_1

of nuclear bombs, a technological devastation that can be avoided by political action, but by a series of complex threats to the fragile ecosystem on which all our lives depend. These threats cannot be so easily avoided and, indeed, even the most radical and decisive action taken immediately could not avoid the reality that we have already altered the climate and destroyed many species with consequences we don't fully understand. What is new about this situation is, firstly, that we are facing grave threats of a kind humanity has never before experienced and, secondly, that the origin of these threats derives from key elements of the ways in which we organise and provision our societies, particularly their high levels of dependence on energy much of it generated through fossil fuels. As the United Nations Development Programme (UNDP) put it succinctly: 'Our development model is bumping up against concrete limits' (UNDP 2011: 15).

Though awareness of the dangers posed by climate change has been growing over recent years, informed by the increasing urgency expressed in the 4th and 5th assessment reports of the Intergovernmental Panel on Climate Change in 2007 and 2013/2014 respectively (IPCC 2007a, b, c, d, 2013, 2014a, b, c), public perception of its gravity has been manipulated and diluted by the activities of climate deniers (Jacques 2012). As a result, hugely disproportionate media attention has been given to individuals and organisations with little or no relevant expertise, making fallacious and inaccurate statements. These often challenge altogether the now well accepted scientific evidence that global temperatures are rising or that it is anthropogenically caused, or minimise the significance of its impacts and exaggerate the costs of its remedy. This deeply corrupted practice is strikingly similar to that which occurred with the link between smoking and cancer. Involving many of the same organisations, and using the same tactics, the aim is to keep the controversy alive by spreading doubt and confusion among the public when the scientific debate has already been sufficiently settled (Oreskes and Conway 2012). However, despite this distraction, the science is fully accepted among governments, scientists and science institutions and the public, policy and technical discourse is indeed evolving. The science of climate change, both in terms of understanding the unequivocal statement in the 5th assessment report of the IPCC that 'human influence on the climate system is clear, and recent anthropogenic emissions of greenhouse gases are the highest in history' (IPCC 2014c: 3) has helped move the debate into a sharper focus on the dangers posed to human civilisation and a recognition, as expressed in the

2015 Paris Agreement, of the important role that 'sustainable lifestyles and sustainable patterns of consumption and production' must play in addressing climate change (UNFCCC 2015: 20).

The debate on climate change is now moving from having a predominant focus on techno-economic means to reduce greenhouse gas (GHG) emissions to a focus on achieving a low-carbon society by 2050. Such a focus is at last consistent with the recognition that it is lifestyles, particularly of the affluent (such as the patterns of consumption and mobility) and the forms of social organisation (industrial scale production of goods, including food, and the governance structures that support them) that require radical change; technology offers some of the tools to effect the necessary change but it cannot address all of the necessary drivers and such tools cannot be divorced from the social context in which they are developed and implemented. Yet, as the focus moves from technology to society, large new debates are beginning to open up related to pathways[1] to a low-carbon society. This is the subject matter of this book. To set the context, this first chapter moves in its next section to outlining the complex dimensions of the problem we face before then examining the dominant responses that have emerged and their inadequacy to the scale of the problem. The subsequent section will analyse the tension between scientific evidence and socio-political ideology that characterises the disjuncture between the scale of the problems being faced and the meagre responses being given. The final section outlines the rest of the book, focusing on the nature of the 'profound shift' now facing society throughout the world.

A 'Wicked Problem' of Many Dimensions

Though we talk about the problem of 'climate change', this is in effect shorthand for a much larger set of interconnected issues that pose major challenges for society worldwide, of which changes in climate are just one manifestation. Different aspects have caused concern at different periods since the nineteenth century but together they constitute what social scientists often call a 'wicked problem', namely, one that resists definition and is not amenable to resolution. The label 'environmental' offers a category that encompasses the many dimensions of the problem but offers little by way of diagnosis or prescription.

While 'environmentalism' as a social movement is dated back to the 1960s, modern Western concerns about environmental limits and the need for conservation of nature and wildlife date back to the second half of the nineteenth century with the establishment of conservation organisations in Britain and attempts at conservation in European colonies. For example, a conference of European powers with colonies in Africa (Britain, Germany, France, Portugal, Spain, Italy and Belgium) met in London in 1900 to sign a Convention for the Preservation of Animals, Birds and Fish in Africa while an International Congress for the Preservation of Nature was held in Paris in 1909 (Adams 2009: 31–33). Contemporary concerns with loss of biodiversity can be traced back to these efforts through such organisations as the International Office for the Protection of Nature (IOPN: 1934), the International Union for the Protection of Nature (IUPN: 1948) and the World Wildlife Fund (WWF: 1961).

In conjunction with concerns about conservation, the science of ecology was developed to analyse patterns of change in natural systems and the impact of human societies upon these. From this the concept of the ecosystem emerged, now much used in contemporary discourse, but it was understood in a more technocratic way relating to the management of nature (Botkin 1990). Ecology helped inform development thinking, alerting to the effects of development on the environment and formulating principles of environmental impact assessment to manage them. An early application of these in the 1960s was in the building of dams. However, beyond the technocratic concern with avoiding the worst effects of development on the natural environment emerged two major concerns that related more centrally and in a more challenging way to features of the dominant model of development.

One was what Paul Ehrlich called 'the population bomb', the title of his book which warned that population growth was going to outstrip the capacity of nature to support it and result in mass starvation (Ehrlich 1972). Often called neo-Malthusian after Thomas Robert Malthus whose 1798 essay on population predicted that its growth would eventually outstrip food supply, these concerns with population have receded in prominence. However, influential authors like James Lovelock, who coined the Gaia hypothesis of the Earth as a complex interactive living system functioning as a single organism, argues that the Earth's present population 'is wholly unsustainable' and that 'we would be wise to aim at a stabilized population of about half to one billion' (Lovelock 2007: 181–182).

The other relates to growth. The influential book *Limits to Growth* published in 1972 by a team from the Massachusetts Institute of Technology for the Club of Rome, sought to model the consequences of complex interactions between the human and planetary systems and predicted overshoot and collapse of the global system by the mid to late twenty-first century on current trends (Meadows et al. 1972). Widely criticised and dismissed at the time, the book was updated after 30 years (Meadows et al. 2004) and again in 2012 (Randers 2012) and its concern with how exponential growth interacts with finite resources has motivated a growing literature more recently (Jackson 2009; Latouche 2009; Heinberg 2011; D'Alisa et al. 2015). Heinberg makes the case as follows: 'From now on, only relative growth is possible: the global economy is playing a zero-sum game, with an ever-shrinking pot to be divided among the winners' (2011: 2). This means planning a transition from a growth-based economy to 'a no-growth economy' or 'a healthy equilibrium economy' (ibid.: 21).

Before introducing more contemporary concerns with climate change and greenhouse gas (GHG) emissions, there is one more conceptual development to introduce. This is the concept of 'sustainable development' that emerged from the World Commission on Environment and Development established by the UN General Assembly in 1983, chaired by the Norwegian Prime Minister, Gro Harlem Brundtland. Its report, entitled *Our Common Future*, was published in 1987 (Brundtland 1987). This sought to square the circle between development, based on economic growth, and environmental limits. However, these latter are not set by the environment but rather by technology and social organisation which could help to ensure that growth both lifts people out of poverty but at the same time conserves and enhances the resource base on which development depends. The concept became so influential that it 'now commands authoritative status, acting as a guiding principle of economic and social development' though 'those that have engaged with the promotion of sustainable development have not adhered to all its principles or its recommended practices' (Baker 2006: 218).

Probably the only dimension of the problem we are discussing that does not fit the definition of a 'wicked problem' is the discovery, in 1985 by the British Antarctic Survey, of a thinning in the ozone layer and a hole in springtime over the Antarctic area. This layer or shield in the Earth's stratosphere absorbs most of the Sun's ultraviolet radiation so that its erosion poses potentially serious damage to humans and other life forms. However, the source of the problem was identified as deriving from ozone-depleting substances, particularly chlorofluorocarbons (CFCs) used in the manufacture of refrigerators. This meant that it was amenable to global political action which resulted in the 1987 Montreal Protocol on Substances that Deplete the Ozone Layer, strengthened in 1990 to require the phasing out of CFCs and other ozone-depleting chemicals by 2000. These actions proved successful in eliminating the source of the problem and allowing the ozone layer to strengthen, a success often contrasted with the failure of international politics substantially to reduce the emissions of GHGs.

Given the urgency now associated with GHG emissions, what is surprising is just how recently it has emerged as a major political issue. While the science of climate change goes back to the Irish physicist John Tyndall (1820–1893), the French mathematician, Joseph Fourier (1768–1830) and the Swedish chemist, Svante August Arrhenius (1859–1927), each of whom postulated various parts of the problem, it remained a low-level concern politically.[2] Indeed, in the 1970s, public concern grew of 'global cooling' and the risk of a new ice age[3] despite firm evidence emerging from a number of independent lines of research pointing to a future warming of the planet.[4] In 1958 Thomas Keeling began measuring the concentration of atmospheric CO_2 and the growing trend shown by these measurements has become a vivid illustration of the reality of carbon emissions. In a short space of time global warming came to be recognised from the annals of the science to the halls of public discourse. As Maslin puts it: 'By the late 1980s, the global annual mean temperature curve rose so steeply that all the dormant evidence from the late 1950s and 1960s was given prominence and the global warming theory was in full swing' (Maslin 2014: 16).[5] This reflects significant advances in global climate modelling, advances that have continued to the present day and are reflected in the ever more firm evidence produced in the reports of the Intergovernmental Panel on Climate Change (IPCC). The growing alarm being expressed in these reports has now brought the issue of GHG emissions to the centre of global politics, as expressed in the climate summit in Paris in December 2015. The latest evidence, as reported in the 2014 report of the IPCC, is summarised in Box 1.1.

Box 1.1 Unprecedented Changes over Decades to Millennia
The 5th assessment report of the Intergovernmental Panel on Climate Change is the latest and most authoritative (because it is the most broad-ranging) of the five reports produced by the IPCC since it was founded in 1988 (1990, 1996, 2001, 2007 and 2013/14). It summarises the latest findings on climate change and its impacts.

Observed Changes in the Climate System

'Warming of the climate system is unequivocal, and since the 1950s, many of the observed changes are unprecedented over decades to millennia. The atmosphere and ocean have warmed, the amounts of snow and ice have diminished, and sea level has risen' (IPCC 2014c: 40).

- *Atmosphere*: 'Each of the last three decades has been successively warmer at the Earth's surface than any preceding decade since 1850' (40).
- *Oceans*: 'Ocean warming dominates the increase in energy stored in the climate system, accounting for more than 90% of the energy accumulated between 1971 and 2010 with only about 1% stored in the atmosphere' (40).
- *Ice and snow*: 'Over the last two decades, the Greenland and Antarctic ice sheets have been losing mass. Glaciers have continued to shrink almost worldwide (high confidence). Northern Hemisphere spring snow cover has continued to decrease in extent (high confidence)' (42).
- *Sea level*: 'Over the period 1901–2010, global mean sea level rose by 0.19 [0.17–0.21] m. The rate of sea level rise since the mid-nineteenth century has been larger than the mean rate during the previous two millennia' (42).

Drivers

'Anthropogenic greenhouse gas emissions have increased since the pre-industrial era driven largely by economic and population growth. From 2000 to 2010 emissions were the highest in history. Historical emissions have driven atmospheric concentrations of carbon dioxide, methane and nitrous oxide to levels that are unprecedented in at least the last 800,000 years, leading to an uptake of energy by the climate system' (44).

- *GHG concentrations*: 'Atmospheric concentrations of GHGs are at levels that are unprecedented in at least 800,000 years. Concentrations of carbon dioxide (CO_2), methane (CH_4) and nitrous oxide (N_2O) have all shown large increases since 1750 (40%, 150% and 20%, respectively)' (44).
- *Human activities*: 'About half of the cumulative anthropogenic CO_2 emissions between 1750 and 2011 have occurred in the last 40 years' (45).

Attribution of Climate Change Impacts

'The evidence for human influence on the climate system has grown since AR4 [previous IPCC report in 2007]. Human influence has been detected in warming of the atmosphere and the ocean, in changes in the global water cycle, in reductions in snow and ice, and in global mean sea level rise; and it is extremely likely to have been the dominant cause of the observed warming since the mid-twentieth century. In recent decades, changes in climate have caused impacts on natural and human systems on all continents and across the oceans. Impacts are due to observed climate change, irrespective of its cause, indicating the sensitivity of natural and human systems to changing climate' (47).

- Human influence: 'It is extremely likely that more than half of the observed increase in global average surface temperature from 1951 to 2010 was caused by the anthropogenic increase in GHG concentrations and other anthropogenic forcings together' (48).
- Observed impacts: 'In recent decades, changes in climate have caused impacts on natural and human systems on all continents and across the oceans. Impacts are due to observed climate change, irrespective of its cause, indicating the sensitivity of natural and human systems to changing climate' (49).

Extreme Events

'Changes in many extreme weather and climate events have been observed since about 1950. Some of these changes have been linked to human influences, including a decrease in cold temperature extremes, an increase in warm temperature extremes, an increase in extreme high sea levels and an increase in the number of heavy precipitation events in a number of regions' (53).

Exposure and Vulnerability

'The character and severity of impacts from climate change and extreme events emerge from risk that depends not only on climate-related hazards but also on exposure (people and assets at risk) and vulnerability (susceptibility to harm) of human and natural systems' (54).

The evidence of changes happening that are unprecedented over a long time period, are increasing in intensity and severity, are more and more caused by human activities, and are exacerbating risks to humans and to nature is now overwhelming. Never in human history has a more comprehensive and unequivocal consensus been reached on such a complex and hazardous environmental phenomenon. As research capability and methods improve, these findings are being refined and strengthened all the time.

No one term is therefore adequate to express the reality of the problem being faced. Perhaps the closest we have come is in the work of the Stockholm Resilience Centre which has issued two reports that seek to define planetary boundaries (PB). The first, published in 2009, sought to identify nine global priorities relating to human-induced changes to the environment. As the Centre states on its website (stockholmresilience.org): 'The science shows that these nine processes and systems regulate the stability and resilience of the Earth System—the interactions of land, ocean, atmosphere and life that together provide conditions upon which our societies depend.' In 2009, three of these boundaries had already been crossed; by 2015 a fourth had been. Furthermore, the research identifies two of these boundaries as being core boundaries, meaning that each 'has the potential on its own to drive the Earth System into a new state should they be substantially and persistently transgressed' (Steffen et al. 2015: 1). These are climate change and biosphere integrity (biodiversity loss and species extinction): by 2015 both had been substantially transgressed. Of the other seven, biogeochemical flows (phosphorus and nitrogen cycles) and land-system change (such as forest cover) had also transgressed planetary boundaries. Of the remainder, ocean acidification is worsening and nearing the boundary, freshwater use is comfortably within boundaries, while global-level boundaries for two processes cannot yet be quantified: these are aerosol loading (microscopic particles that affect climate and living organisms) and novel entities (such as organic pollutants, radioactive materials, nanomaterials, and micro-plastics). Only one boundary shows evidence that human actions have helped keep it safely within planetary boundaries, that is stratospheric ozone depletion.

The boundaries identified by the Stockholm Resilience Centre detail the multidimensional nature of the challenges now faced by humanity. It is clear from the brief description above that the drivers of the processes that are pushing society to transgress boundaries derive from products and activities that are core features of modern society; some of them have elicited concern for over 150 years but actions taken have not been sufficient to keep them safely within the carrying capacity of the planet. This has already been determined clearly by the IPCC in terms of climate change. Action on ozone is the only exception. As the authors of the 2015 report on Planetary Boundaries state:

> The precautionary principle suggests that human societies would be unwise to drive the Earth System substantially away from a Holocene-like condition. A continuing trajectory away from the Holocene[6] could lead, with an uncomfortably high probability, to a very different state of the Earth System, one that is likely to be much less hospitable to the development of human societies. (ibid.: 1–2)

This is a clear call for society to change its developmental direction before it is too late.

Inadequate Responses

Six weeks after the ground-breaking Paris climate summit in December 2015, the oil multinational BP released its annual energy outlook in mid-February 2016. Despite oil having fallen over the previous months to a fraction of its price and BP having to lay off thousands of workers as a result, the report forecast that we are on the brink of a new oil and gas boom, led by US shale oil. Furthermore, with the world's states pledging themselves in the Paris Agreement to keep global warming to well below 2° Celsius and with a growing movement to divest from oil companies, BP forecast that non-oil transport would grow by only 5% over the following two decades and that demand for oil and gas will continue to grow, particularly in the developing world. As Bob Dudley, group chief executive wrote: 'There are clear signs that the market is adjusting and that it will gradually rebalance.' He foresees that fossil fuels will supply 60% of the energy increase up to 2035 and that carbon emissions are likely to further increase requiring more policy action. While responding to immediate challenges 'we mustn't lose sight of the longer-term role of our industry in providing the energy the world needs to grow and prosper, and doing so in a safe and sustainable manner' (BP 2016: 4, 5).

Neither are BP alone. Carbon Tracker, the not-for-profit financial think tank that seeks to align capital market actions on global energy with climate realities, reports that BP projects a 24% increase in fossil fuel use by 2035 but that Exxon expects a 27% fossil fuel increase, Shell a 37% increase and OPEC a 54% increase, all by 2040 (Campanale 2016). What Carbon Tracker draws attention to is the danger of 'stranded assets', namely investments that will not provide a return as the world moves to renewable energy sources in order to live within carbon budgets. It concludes: 'Looking at performance in recent years, the companies have been spending more just to stay still in terms of production and have not seen growth in volumes. It is therefore unsurprising that with prices more volatile, and capex [capital expenditure need for investment] cut, the companies will struggle to maintain volumes' (Carbon Tracker 2015: 23). These stranded assets it estimates as being worth as much as $2.2 trillion, both existing investments in the oil, gas and coal industries and the amounts companies will be seeking in order to maintain investment over the next decade. For Carbon Tracker, it makes financial sense for investors to move away from fossil fuel companies and into renewables and this call is having an impact. It was announced at the Paris climate summit that by then some 500 institutions representing more than $3.4 trillion in assets had made some form of commitment to divest from fossil fuel companies, including local authorities, pension funds, companies, churches and universities; and it has sparked a global mass movement for divestment.

However, what is illustrated by this example is that BAU (business as usual) continues to dominate how major institutions see and plan for the future (see Box 1.2). A number of leading international organisations have drawn attention to this disjuncture between the scientific evidence and the socioeconomic trajectory of today's world. The World Bank, normally a cautious and even conservative organisation, issued a series of three reports in 2012, 2013 and 2014 under the title *Turn Down the Heat*. Prepared by the Potsdam Institute for Climate Impact Research and Climate Analytics, these warned that the world is on track to warm by 4°C above pre-industrial times by the end of the twenty-first century with devastating consequences including coastal cities being under water, high rates of malnutrition, unprecedented heat waves especially in the tropics, severe water scarcity in many regions, more intense tropical cyclones with severe destructive force and extensive irreversible loss of biodiversity (World Bank 2012). Based on the 5th assessment report of the IPCC (2013, 2014a, b, c), the World Bank concludes that 'climate policy has not to date succeeded in curbing global greenhouse gas emissions, and emissions are steadily rising' but that this 'does not imply a lock-in to a high-emitting pathway if there is a move toward rapid, technically and economically feasible mitigation' (World Bank 2014: 6). In his foreword to the third report, the President of the World Bank, Jim Yong Kim, wrote that the findings on the impacts of global warming around the world are 'alarming' and holding warming below 2°C 'will require substantial technological, economic, institutional and behavioural change. It will require leadership at every level of society' (ibid.: xiii and xiv).

Box 1.2 Hard-Wired to Avoid the Threat of Climate Change?

We still don't treat climate change with the reverence we reserve for something like a terrorist attack, writes *Guardian* columnist Ruth Greenspan Bell. Maybe the blame for our inaction lies with our very nature: 'Evolution did not design our bodies to treat climate change with urgency'. She argues that we are hard-wired for immediate real-time threats over those taking place on an extended time scale, as is climate change.

'The challenge in moving more forcefully to stop the flow of greenhouse gases is that if you have to stop and think about whether a specific action or activity is threatening, that very process engages very different parts of the human brain, and not the ones that impel us to action', she writes. The hormones that provide us with increased strength and speed don't kick in when we have to go and research to understand the seriousness of the threat we face.

The result is that we pay attention to climate change only occasionally, when a natural disaster hits that affects us or when we are bombarded by saturation media coverage of gatherings like the Earth summit in Rio in 1992 or the Paris climate change summit of December 2015. But these moments pass and we are inclined to leave it to the experts whom we believe are taking care of it for us (Greenspan Bell 2016).

A report from the OECD in 2011 showed how 'the world is locking itself into high carbon systems more strongly every year' as demand for cars grows in developing countries, as growing energy demand world-wide outstrips the capacity to improve the energy intensity of economies, and as expanding agricultural areas most particularly in Africa increase land-use emissions. The report states: 'Most countries have begun to respond through actions at the international, national and local levels, drawing on a mix of policy instruments that include carbon pricing, other energy-efficiency policies, information-based approaches and innovation. Some progress can be noted, but much more needs to be done to achieve the 2°C goal' (OECD 2011: 5–9). In a report prepared for the Paris climate summit and assuming no additional mitigation measures are taken beyond those already adopted, the OECD estimates that global GDP will decline by between 1% and 3.3% by 2060; but as temperatures continue to rise global GDP could fall by up to 10% by the end of the century. Changes in crop yields and in labour productivity are likely to have the most impact while damage from sea level rise will become gradually more important, especially after mid-century. Net economic consequences are likely to be negative in 23 of the 25 regions used in the OECD modelling. The report adds that the costs of inaction 'likely underestimate the full costs of climate action impacts' (OECD 2015: 13). Early and ambitious action can help to reduce the long-term costs, it adds.

In highlighting the 'concrete limits' facing our current develop model, the UNDP take a much broader focus than economic growth. They point to the scale of the challenges when recognising the 'fundamental contradiction: business-as-usual is neither sustainable nor equitable' (UNDP 2011: 82). Instead, what is required is 'a fundamental rethinking of the conventional growth model', including 'an expansive rethinking of the role of the state and communities' (ibid.: 81). This will require more active public policy towards decoupling development from carbon emissions and incorporating the true value of ecosystem services into national development plans. It goes on:

> A key constraint to public action on environmental problems is lack of awareness. About a third of the world's people seem unaware of climate change, and only about half consider it a serious threat or know that it is caused at least partly by human activity. But even with raised awareness, serious political constraints would remain—in other words, our collective failure

> to act also reflects the complexity of the politics and the power of groups opposing change. ... [M]any countries and communities most affected by climate change lack power and influence. So understanding these constraints is a vital first stop in framing strategies with a real chance of meaningful change. (ibid.: 82–83)

For this reason, the UNDP places emphasis both on empowering people to bring about change and on reforming institutions so that they help enable such change (a fair and independent judiciary and the right to information from government are mentioned). Civil society organisations have an important role to play in helping empower citizens and put pressure on governments to be more responsive.

The UNDP's analysis brings power inequalities into focus, thereby making explicit what was illustrated in the example of oil company projections and plans with which this section opened. For there we see pitted against one another two major power blocs that exercise great influence in configuring the shape of our carbon-intensive model of development, namely the fossil fuel industry and the financial sector that up to now has supported it. These factors are too often left out of the analysis of why our global society has been so unable to take the steps necessary to address the challenges of climate change and its potentially devastating impacts on our world and our livelihoods. Yet, despite more than two decades of action at global level through the United Nations Framework Convention on Climate Change (UNFCCC) and its impact on national policy and practice throughout the world, and the growing evidence of the grave threat of climate change detailed in the reports of the IPCC, there is a widespread consensus that the actions taken so far have been completely inadequate.

It is this disjuncture between the scale of the crisis we face and the poverty of responses to it that is one of the most serious problems to be addressed. However, very little attention has been paid to the reasons for this; indeed, it seems to be taken for granted as if it were somehow to be expected. But it is noteworthy that a society that likes to think of itself as scientific and as making policy based on evidence, when faced with overwhelming evidence of the gravest threats to its future, takes minimal measures to avoid these. Perhaps the most insightful analysis of the reasons for this, at least at an authoritative global level, comes from a most surprising source. Pope Francis, in his encyclical letter *Laudato Si: On Care for our Common Home*, recognises that 'we lack leadership capable of striking out on new paths and meeting the needs of the present with concern for all and without prejudice towards coming generations' (Pope Francis 2015: par. 53). He goes on to identify the source of this weakness:

> It is remarkable how weak international political responses have been. The failure of global summits on the environment make it plain that our politics are subject to technology and finance. There are too many special interests, and economic interests easily end up trumping the common good and manipulating information so that their own plans will not be affected. The Aparecida Document [of the Latin American Bishops' Conference in 2007] urges that 'the interests of economic groups which irrationally demolish sources of life should not prevail in dealing with natural resources'. The alliance between the economy and technology ends up sidelining anything unrelated to its immediate interests. Consequently the most one can expect is superficial rhetoric, sporadic acts of philanthropy and perfunctory expressions of concern for the environment, whereas any genuine attempt by groups within society to introduce change is viewed as a nuisance based on romantic illusions or an obstacle to be circumvented. (par. 54)

This identifies special interests associated with technology and the economy as sidelining efforts to address adequately and imaginatively the threats faced by humanity and to change our developmental direction. Can it be that ideology and special interests are sidelining the evidence of science?

Evidence Versus Ideology

So, the evidence is clear but the actions taken so far have been minimalist, and inadequate to reverse GHG emissions substantially or to address the other aspects of our development that result in surpassing planetary boundaries. As the UNDP put it, understanding constraints to action is a vital first step to framing strategies to address the problem. A helpful beginning is to follow Pope Francis in identifying 'the alliance between the economy and technology [which] ends up sidelining anything unrelated to its immediate interests'. If this is true then it constitutes a major obstacle to decisive action. But how true is it?

Certainly, technological solutions figure prominently in the literature on climate change. For example, climate scientist Mike Hulme details a range of what he calls 'techno-fixes' including stratospheric aerosol injection, marine cloud brightening, orbital mirrors, urban whitewashing, biochar, ocean fertilisation, carbon capture and storage and enhanced weathering (Hulme 2014: 7–10). As Hulme points out, these are often presented as alternatives to the political efforts to reduce emissions; the latter's failure, it is argued, justifies and even necessitates technological solutions (ibid.: 14). Yet, he makes two critiques that point to the perennial dangers of placing our hope in science: firstly, 'it suggests a supreme confidence in human knowledge and ingenuity—a confidence approaching arrogance'; secondly, it 'reveals a certain poverty of the imagination, a preference for technical calculus that has little regard for the relational, creative and spiritual dimensions of *anthropos*, the human being' (ibid.: 111; italics in original). Yet, in positing this neat opposition between the technological and the human, does Hulme underestimate the extent to which technology has colonised human society and constrained the human imagination within technological boundaries? It is interesting that the IPCC, in its 5th assessment report, draws attention to the high costs involved and the potential risks or 'pervasive uncertainties involved in nearly all techniques'. It reports the argument that 'geoengineering could become a distraction from urgent mitigation and adaptation measures' (Clarke et al. 2014: 484).

The French philosopher of the socio-anthropology of technology Alain Gras reminds us that a technical object is always part of a technical system (Gras 2017: 10) rather than being simply a set of neutral tools. As Gras puts it: 'This new Anthropocentric society that relies on the power of heat in machines is nothing but a *thermo-industrial civilization*', involving 'a real upheaval in the representation of the relationship of humans with the world around them' (ibid.: 13; italics in original). Central to this civilisation as it evolved following the industrial revolution were 'technical macro-systems of great complexity in which the railway network was the original model followed by the electrical network' and requiring a huge social infrastructure of railways, paved roads, energy grids, oil pipelines, ships and tanks, based on what Gras calls 'the excessive use of fossil fuels' (ibid.: 21), finally generating today's globalisation. Therefore the 'technical calculus' mentioned by Hulme in the previous paragraph turns out to be the dominant response to climate change emerging from the heart of the technological civilisation to which the industrial revolution has given rise. As Gras points out, 'we are locking ourselves into an electronic world' (ibid.: 26) which is both functional to the needs of neo-liberalism but also creating a world of surveillance of the private sphere 'in a more insidious way than an old-style totalitarian regime would have been able to do' (ibid.: 26). An energy transition based on renewables is therefore a 'fake' change masking a real continuity in technological civilisation. Electricity increases the destructive potential of the world, he argues, 'via the new territorial focus of predation—digging activity—it gives rise to. Lithium for batteries will wipe out Salar de Uyuni in Bolivia, cobalt and coltan are destroying the Congo (Kivu, Katanga), and different rare materials for wind turbines (neodymium, which replaces copper), as well as all those rare materials needed by the sophisticated tools of the pseudo 'energy transition', render a totally illusory green transition' (ibid.: 27).

Gras's analysis has the advantage of revealing the extent to which technology shapes our world—both in material terms but also through shaping our dominant responses to problems like climate change—and situating this within the dominant ideology of a growth-based globalised neo-liberalism. Pope Francis's 'alliance between the economy and technology' turns out therefore to identify very precisely the principal way in which our responses to climate change are configured, and it is so dominant we are largely unaware of it. In discussing what he calls 'the technological paradigm', the Pope further echoes the themes developed by Gras:

> It can be said that many problems of today's world stem from the tendency, at times unconscious, to make the method and aims of science and technology an epistemological paradigm which shapes the lives of individuals and the workings of society. The effects of imposing this model on reality as a whole, human and social, are seen in the deterioration of the environment, but this is just one sign of a reductionism which affects every aspect of human and social life. We have to accept that technological products are not neutral, for they create a framework which ends up conditioning lifestyles and shaping social possibilities along the lines dictated by the interests of certain powerful groups. Decisions which may seem purely instrumental are in reality decisions about the kind of society we want to build. (par. 107)

This paradigm, which has now been globalised, he links to the issue of economic growth, naming some of the interest groups that promote it. The relationship between human beings and material objects, he writes, 'has made it easy to accept the idea of infinite or unlimited growth, which proves so attractive to economists, financiers and experts in technology. It is based on the lie that there is an infinite supply of the earth's goods, and this leads to the planet being squeezed dry beyond every limit' (par. 106), a veiled reference to the issue of planetary boundaries. Furthermore, this technocratic paradigm 'tends to dominate economic and political life', he writes (par. 109) and 'tends to absorb everything into its ironclad logic', so that 'it has become countercultural to choose a lifestyle whose goals are even partly independent of technology, of its costs and its power to globalise and make us all the same' (par. 108). This therefore is a major obstacle to facing the deeper challenges raised by climate change: 'To seek only a technical remedy to each environmental problem which comes up is to separate what is in reality interconnected and to mask the true and deepest problems of the global system' (par. 111). It identifies the major ideological barrier to understanding the full import of the challenges we face and to fashioning responses adequate to the evidence of science.

In facing the scale of the transition to move to a low or post-carbon society,[7] going beyond the limits of the technological paradigm, analysts have returned to the writings of Karl Polanyi (1886–1964) whose classic book entitled *The Great Transformation* was published in 1944 (Polanyi 2001) (Box 1.3). This book is a history of the British Industrial Revolution, interpreting it as a 'utopian experiment' involving the imposition on society of the self-regulating market and leading to the creation of 'a market society' (Polanyi 2001: 258). This required treating land, labour and money as commodities, bought and sold on the market, what Polanyi called 'fictitious commodities' which 'means no less than the running of society as an adjunct to the market' (ibid.: 60). According to Polanyi, 'the commodity fiction disregarded the fact that leaving the fate of soil and people to the market would be tantamount to annihilating them' (ibid.: 137) and he argued that the end result was that the system 'required that the individual respect economic law even if it happened to destroy him' (ibid.: 89). Inevitably society reacted against this destructive imposition resulting in what Polanyi called the 'double movement', the first movement being imposing the market on society while the second movement was the spontaneous reaction of society against this imposition.

Box 1.3 The Great Transformation: Turning to Karl Polanyi

The German Advisory Council on Global Change (WBGU), an independent scientific advisory body to the German government established in 1992, compares the transition to a low-carbon society to 'the two fundamental transformations in the world's history:

the Neolithic Revolution, i.e. the invention and the spread of farming and animal husbandry, and the Industrial Revolution' (WBGU 2011: 5). It makes clear that the most difficult changes required for this transition 'transcend technologies—changing lifestyles, for instance, or revolutionising global cooperation, overcoming policy-related barriers, and dealing responsibly with permanent, cross-generational changes' (ibid.: 82). The Council describes this as the Great Transformation, 'not least with reference to Karl Polanyi's "Great Transformation" (1944) to describe an all-encompassing transition' (ibid.: 83).

When the New Economics Foundation, a British independent think-tank on economic wellbeing, wrote a report on how to redesign the economic system to overcome a series of major challenges by 2050, it called the process the Great Transition 'as a deliberate echo' of Polanyi's *The Great Transformation*. The interlinked systemic problems it identified were consuming beyond planetary limits, untenable inequality, growing economic instability and a breakdown in the relationship between 'more' and 'better'. Executive director Stewart Wallis writes in the Foreword that Polanyi analysed how market processes in the industrial revolution 'created severe ruptures in the fabric of social life' and argued that what was needed was 'to find a balance between the market and the non-market; the private and the public; the individual and the community'. The need for this balance is all the more pressing today given the huge environmental problems we face, states Wallis (NEF 2009: 1).

Three aspects of Polanyi's analytical framework are important for this discussion. The first refers to the impact of the marketisation of society on social thought. 'Nothing obscures our social vision as effectively as the economistic prejudice', he warned (ibid.: 166), namely the influence of market processes on obscuring a full appreciation of their destructive impact on society and on the ecosystem. This can be seen as another key dimension of the dominant epistemological paradigm referred to by Pope Francis. The second refers to the commodification of land, namely treating nature as a series of commodities to be bought and sold. Polanyi was prescient in identifying how this inevitably leads to ecological destruction, and it alerts us to the dangers of using market mechanisms as solutions to environmental problems as is so common today (carbon pricing and markets). The third aspect links back to the topic of technology already discussed. For Polanyi was gravely concerned at the 'subordination of man to the needs of the machine': 'behind the fading fabric of competitive capitalism there looms the portent of an industrial civilisation, with its paralysing division of labour, standardisation of life, supremacy of mechanism over organism, and organisation over spontaneity. Science itself is haunted by insanity. This is the abiding concern', he wrote in 1947 (Polanyi 1968: 59).

We can identify then in the technological paradigm and the economistic prejudice the principal constraints to addressing adequately the multi-dimensional problem of a development model that is surpassing the carrying capacity of the biosystem and putting the future wellbeing of humanity on the planet at grave risk. Instead what we have are actions that either prioritise technological fixes to the problems and/or that are constrained by having to conform to the needs of a neo-liberal and globalised market system. Indeed, there is a strong tendency to adopt actions that are seen as fuelling the development of that system through offering new 'green' investment opportunities. Yet, all of these have so far been completely inadequate to shift human society towards a more regenerative relationship with the ecosystem on which it depends but on which it is imposing severe strain. Making the shift towards this sort of relationship is now the single most urgent task for politics and society at all levels, from global to local.

Starting 'A Profound Shift'

When addressing the Paris climate summit in 2015, the host President Hollande stated:

> I am going to be frank: to resolve the climate crisis, goodwill and statements of intent are not enough. We are coming to a breaking point. Paris must be the starting point of a profound shift. We can no longer consider nature as a mere bottomless reservoir of resources there for our sole and full benefit. This transformation is both a moral obligation and a global opportunity, for it opens up possibilities for development with renewable energies, clean transport, waste recycling, agro-ecology, preservation of biodiversity, and universal access to all global public goods.

A profound shift to a low-carbon society requires paying attention to the fact that change in all societies happens within development pathways, whether explicitly defined or not, and that the nature of these pathways fundamentally shapes the outcomes achieved. Furthermore, such pathways are by definition richly multifaceted, including not just economic development and technological innovation but social practices, governance systems, cultural values and historical factors such as path dependencies, power relations and social structures. As O'Mahony and Dufour put it:

> Development pathways can emphasise the multiplicity of possible outcomes, the multidimensional problem character and the power of human agency. For a development path to be sustainable in the long term wealth, resources, and opportunity must be shared for minimum standards of security, human rights, and social benefits, such as food, health, education, shelter and the opportunity for self-development. (O'Mahony and Dufour 2015: 416)

Though a whole field of study grew up around development models in the post-War period (see Kirby 1997 for a survey), this field of study has been little applied to the challenges of climate change. This may be because, like most of the social sciences (with the partial exception of geography as a discipline), it has never integrated the ecosystem as an essential context within its conceptual toolkit. If development studies has tools of analysis to contribute to mapping out pathways towards a low-carbon society, then it is equally true that climate change also challenges development to begin to nest itself within planetary boundaries.

This book draws on the rich terrain of development theory, and in particular on the international political economy of development, to engage with the field of global scenario studies studying the interaction of variables such as population, energy and GHG emissions in the context of the need to transition to a low-carbon society. It has been recognised that this field has difficulties in dealing with the complexity of economic and technological variables, but also, more generally, with 'uncertainty and driving forces which may not be quantified, primarily social, cultural and governance' (O'Mahony and Dufour 2015: 415). This book is therefore written by two scholars, one from each field, and is itself an exploration of what can be learnt for pathways towards a low-carbon future from the interaction of both fields of study.

It is divided into four parts. Part I, entitled 'Climate Change as Problem' contains three chapters. The first chapter has introduced the topic, while

Chap. 2 examines the principal ways in which the 'climate change' challenge is conceived of at the moment. The third chapter focuses on policy options, critiquing the inadequacy of the approach outlined in the previous chapter for getting us to a low-carbon society by 2050 in two ways: firstly by examining the scientific evidence and secondly by identifying what is missing, namely development models and all that constitute them—political trajectories, social practices such as consumption, governance systems and dominant values. Part II is on development pathways with Chap. 4 on models, introducing the concept of such pathways and discussing how they coalesce around models with particular state-market-civil society relationships constituting them. Chapter 5 looks at scenarios drawing on insights from scenario analysis as a way of mapping out alternative pathways and helping balance the structural nature of development models by more agency-driven approaches.

Part III examines pathways taken to combine development and sustainability. Chapter 6 examines the types of pathways chosen by a range of countries in planning the transition to a low-carbon society, extracting the lessons to be learnt. It looks at three groups, all of which can be seen as developmental successes: the 'developed world' countries of the US, the EU and Japan, the 'emerging economies' of China, India, Brazil and Mexico, and the Nordics, a group of countries showing leadership on climate change. The following chapter looks at different examples of developing countries, identifying those few that have succeeded in combining development with sustainability, and examining in turn the distinctive challenges facing Latin America, Africa and Asia, giving examples of the variety of responses, from local to national. Attention will be paid to the small island developing states (SIDS), whose very existence is threatened. Again the different ways the challenges are presenting themselves and the different responses being put in place are examined and lessons learnt.

The final part has three chapters examining pathways to a low-carbon future. Chapter 8 analyses the dominant model in place to achieve the transition to a low-carbon society, namely climate capitalism. This uses the mechanisms of capitalism (pricing carbon, trading in carbon credits, offsetting emissions through investing in developing countries, etc.) in an attempt to lower emissions and to create incentives for switching to renewables and for more sustainable lifestyles. Chapter 9 asks whether we need to move to an ecosocialist system given the limitations of climate capitalism, and in particular the emerging debate about moves to a 'steady state economy' or a 'degrowth economy'. This raises the prospect that any successful transition to a low-carbon society will require a new socioeconomic and political system, moving beyond current forms of capitalism and the development paths it has engendered. The final chapter looks at options and prospects. It assesses how far we've been able to go in elucidating a clear pathway through the uncharted and evolving landscape towards a low-carbon future, clarifying the dead ends and the wrong turns, and offering signposts to keep us on the path. The book ends with a hard-headed assessment of the prospects for reaching a low-carbon society by 2050 and what we do if we don't make it.

Notes

1. At times we use 'pathways' in the singular and at times in the plural. The singular usage is to emphasise that whatever ways we develop, they must take us towards a low or post-carbon future. Therefore, all other pathways must be avoided. But the plural usage also acknowledges that there are different mixes of state, market and civil society that potentially can get us there; these include different technological pathways, social pathways, cultural pathways and environmental pathways that are layered in a single 'development pathway'. So we almost certainly will end up with a plurality of pathways adequate to taking us to our destination.
2. President Lyndon B. Johnson's Advisory committee laid out in 1965 that emissions of carbon dioxide from the burning of fossil fuels could rapidly reshape Earth's climate (Revelle et al. 1965).
3. 'Global cooling' had little support in the scientific community but was picked up by the media and led to public confusion that to some extent remains today. The large majority of climate research in the 1970s predicted the Earth would warm as a consequence of CO_2 emissions (for a useful discussion on the history of climate science see Peterson et al. (2008) in the Bulletin of the American Meteorological Society).
4. The first global climate records were established in the 1870s, the first evidence of a global warming trend emerged from analysis in the 1930s, see Peterson et al. (2008).
5. Decades of climate science research in the twentieth century now had voluminous evidence from multiple independent lines of inquiry supporting the theories of climate change that had emerged from the early pioneers of the eighteenth and nineteenth century. The science of climate change had moved from a theory to robust and verified science, and global political and public opinion was on catch-up.

6. The Holocene refers to the geological period that began about 11,000 years ago providing the climatic conditions for the development of human civilisation.
7. While we prefer the term 'post-carbon' society or future, it will be noted that we also use 'low-carbon'. 'Post-carbon' may imply an end globally to the use of fossil fuels over the course of the twenty-first century, and a similar trend for the other greenhouse gas (GHG) emissions from human activities. 'Low-carbon' is the more commonly used term and may imply a very minor role for fossil fuels in the future, of the order <5% of global consumption of coal, oil and gas in 1990 by the year 2100. Both pathways may also require large-scale removal of CO_2 from the atmosphere, see IPCC (2014c).

References

Adams, W.M. 2009. *Green Development: Environment and Sustainability in a Developing World*. London: Routledge.

Baker, Susan. 2006. *Sustainable Development*. London: Routledge.

Botkin, D.B. 1990. *Discordant Harmonies: A New Ecology for the Twenty-first Century*. New York: Oxford University Press.

BP. 2016. *BP Energy Outlook*. Accessed 17 February 2016. http://www.bp.com/content/dam/bp/pdf/energy-economics/energy-outlook-2016/bp-energy-outlook-2016.pdf

Brundtland Commission. 1987. *Our Common Future*. Oxford: Oxford University Press.

Campanale, Mark. 2016. Why Divestment from Fossil Fuels is Now Essential and How Institutional Divestment Has Changed the Debate. Presentation at workshop *Investing Ethically for People and Planet: Ending Irish Investments in Fossil Fuels*, Maynooth University, 16 February 2016.

Carbon Tracker. 2015. *The $2 Trillion Stranded Assets Danger Zone: How Fossil Fuel Companies Risk Destroying Investor Returns*. London: Carbon Tracker Initiative.

Clarke, L., K. Jiang, K. Akimoto, M. Babiker, G. Blanford, K. Fisher-Vanden, J.-C. Hourcade, et al. 2014. Assessing 'Climate Change 2007: Mitigation. Contribution of Working Group III'. *Climate Change and Sustainable Development* 3 (S1): S19–S40.

D'Alisa, Giacomo, Federico Demaria, and Giorgos Kallis, eds. 2015. *Degrowth: A Vocabulary for a New Era*. London: Routledge.

Ehrlich, Paul R. 1972. *The Population Bomb*. London: Ballantine.

Gras, Alain. 2017. The Deadlock of the Thermo-industrial Civilisation: The (Impossible?) Energy Transition in the Anthropocene. In *Transitioning to a Post-Carbon Society: Degrowth, Austerity and Wellbeing*, ed. Ernest Garcia, Mercedes Martinez-Iglesias, and Peadar Kirby, 3–35. Basingstoke: Palgrave Macmillan.

Greenspan Bell, Ruth. 2016. Why Don't We Treat Climate Change with the Rigor We Give to Terror Attacks? *The Guardian*, 15 February 2016.
Heinberg, Richard. 2011. *The End of Growth: Adapting to Our New Economic Reality*. Gabriola Island, BC: New Society Publishers.
Hulme, Mike. 2014. *Can Science Fix Climate Change?* Cambridge: Polity Press.
IPCC. 2007a. *The Physical Science Basis*. Working Group I Report. Geneva: IPCC.
———. 2007b. *Impacts, Adaptation and Vulnerability*. Working Group II Report. Geneva: IPCC.
———. 2007c. *Mitigation of Climate Change*. Working Group III Report. Geneva: IPCC.
———. 2007d. *The AR4 Synthesis Report*. Geneva: IPCC.
———. 2013. *Climate Change 2013: The Physical Science Basis*. Working Group I Report. Geneva: IPCC.
———. 2014a. *Climate Change 2014: Impacts, Adaptation, and Vulnerability*. Working Group II Report. Geneva: IPCC.
———. 2014b. *Climate Change 2014: Mitigation of Climate Change*. Working Group III Report. Geneva: IPCC.
———. 2014c. *Climate Change 2014: Synthesis Report*. Geneva: IPCC.
Jackson, Tim. 2009. *Prosperity Without Growth: Economics Fore a Finite Planet*. London: Earthscan.
Jacques, Peter J. 2012. A General Theory of Climate Denial. *Global Environmental Politics* 12 (2): 9–17.
Kirby, Peadar. 1997. *Poverty Amid Plenty: World and Irish Development Reconsidered*. Dublin: Trócaire and Gill & Macmillan.
Latouche, Serge. 2009. *Farewell to Growth*. Cambridge: Polity.
Lovelock, James. 2007. *The Revenge of Gaia*. London: Penguin.
Maslin, Mark. 2014. *Climate Change: A Very Short Introduction*. Oxford: Oxford University Press.
Meadows, Donella, Dennis Meadows, and Jørgen Randers. 2004. *Limits to Growth: The 30-Year Update*. White River Junction, VT: Chelsea Green.
Meadows, Donella H., Dennis L. Meadows, Jørgen Randers, and William W. Behrens III. 1972. *The Limits to Growth*. New York: Universe Books.
NEF. 2009. *The Great Transition*. London: NEF.
O'Mahony, T., and J. Dufour. 2015. The Social and Cultural Dimensions of Sustainable Development, Mitigation and Scenarios: Grasping the Opportunities for Human Development. In *Sustainable Futures in a Changing Climate*, ed. Aino Hatakka and Jarmo Vehmas. Proceedings of the Conference *Sustainable Futures in a Changing Climate*, 11 and 12 June 2014, Helsinki, Finland. FFRC eBOOK 2/2015. Finland Futures Research Centre, University of Turku, 414–427.
OECD. 2011. *Environmental Outlook to 2050: Climate Change Chapter*. Paris: OECD.

———. 2015. *The Economic Consequences of Climate Change*. Paris: OECD.

Oreskes, Naomi, and Erik M. Conway. 2012. *Merchants of Doubt: How a Handful of Scientists Obscured the Truth on Issues from Tobacco Smoke to Global Warming*. London: Bloomsbury Press.

Peterson, T.C., W.M. Connolley, and J. Fleck. 2008. The Myth of the 1970s Global Cooling Scientific Consensus. *Bulletin of the American Meteorological Society Society* (89): 1325–1337.

Polanyi, Karl. 1968. Our Obsolete Market Mentality. In *Primitive, Archaic and Modern Economies: Essays of Karl Polanyi*, ed. George Dalton, 59–77. New York: Anchor Books.

———. 2001. *The Great Transformation: The Political and Economic Origins of Our Time*. Boston: Beacon Press.

Pope Francis. 2015. *Laudato Si: On Care for Our Common Home*. Vatican City: Vatican Press.

Randers, Jørgen. 2012. *2052: A Global Forecast for the Next Forty Years*. White River Junction, VT: Chelsea Green.

Revelle, R., W. Broecker, H. Craig, C. D. Kneeling, and J. Smagorinsky. 1965. *Restoring the Quality of Our Environment: Report of the Environmental Pollution Panel*. President's Science Advisory Committee, The White House.

Steffen, Will, Katherine Richardson, Johan Rockström, Sarah E. Cornell, Ingo Fetzer, Elena M. Bennett, R. Biggs, et al. 2015. Planetary Boundaries: Guiding Human Development on a Changing Planet. *Science Express*, 15 January 2015.

UNDP. 2011. *Human Development Report 2011: Sustainability and Equity*. New York: Palgrave Macmillan.

UNFCCC. 2015. *Paris Agreement*. Paris: UNFCCC.

WBGU. 2011. *World in Transition: A Social Contract for Sustainability*. Berlin: WBGU.

World Bank. 2012. *Turn Down the Heat: Why a 4° Warmer World Must be Avoided*. Washington, DC: The World Bank.

———. 2013. *Turn Down the Heat: Climate Extremes, Regional Impacts, and the Case for Resilience*. Washington, DC: The World Bank.

———. 2014. *Turn Down the Heat: Confronting the New Climate Normal*. Washington, DC: The World Bank.

CHAPTER 2

Framing the Problem: How the Climate Change Message is Constructed

Introduction

In outlining in Chap. 1 the scale of the challenges faced as a result of humanity living beyond the limits of the planet's ecosystem and the inadequacy of the responses to address this so far, attention has been drawn to our ways of framing the issues involved. This includes forms of discourse that present the challenge in particular ways (summed up in terms like 'climate change', 'conservation', 'environment' and 'sustainability') but it goes beyond simple description in that it incorporates prescriptive and normative elements, namely what should be done to address these challenges. So, for example, a focus on global warming will give priority to curbing GHG emissions, a focus on conservation to protecting habitats and endangered species, and a focus on sustainability to taking environmental concerns into account in socio-economic change. In other words, framing involves not just presenting things in a certain way but it also suggests certain types of responses and detracts from others. Therefore, it is far from neutral or casual and indeed can be said to be quite political, in that it assumes and operates within certain understandings of values, power relations and our relationship to the natural environment, and requires questioning of whether these need to be changed if we are to address climate change challenges successfully.

P. Kirby, T. O'Mahony, *The Political Economy of the Low-Carbon Transition*, International Political Economy Series,
DOI 10.1007/978-3-319-62554-6_2

These insights take us back to the work of the great Italian political writer and activist, Antonio Gramsci (1891–1937). One of Gramsci's major contributions to our understanding of power was his central insight that the dominance of one group over another (the bourgeoisie over the working class in classical Marxist thinking) does not depend so much on the use of force but, rather, on winning the battle of ideas. The group whose ideas and values become the 'common sense' of society do not require force to maintain its hegemony because people have internalised a worldview that supports that hegemony and so they acquiesce in it unthinkingly. This insight has been developed by social constructivists such as Peter L. Berger and Thomas Luckmann (Berger and Luckmann 1991) who argue that norms and beliefs about what constitutes reality become institutionalised in society. As a result, understanding the ideational or discursive construction of a social reality becomes an important task over and above understanding its material manifestations.

For example, feminist scholars have used constructivist approaches in exposing masculinist biases in the ways power is distributed in society thus helping change the social and economic position of women. These approaches have also helped in understanding that economic globalisation cannot be fully understood just by examining material evidence for the intensification of global interconnectedness, since its transformative power owes much to the almost irresistible impact that ideas of economic globalisation have had, particularly among policy makers. McGrew writes that 'in naming or identifying these material trends in the world economy as a process of "economic globalization", that very process becomes socially or discursively constructed and is thus given intersubjective meaning. Social constructivism, therefore, has an important bearing upon how globalisation is interpreted and understood, both within the academy and beyond' (McGrew 2011: 301). This shows that it is important to understand how an issue becomes socially or discursively constructed and the intersubjective meaning it thereby takes on. It is therefore surprising that social constructivism has been little used to examine critically how the topic of climate change has been framed and the implications this holds for dealing with it, highlighting how framing constructs the issue in particular ways that emphasise some aspects but marginalise others, and drawing into the frame the wider ideological and substantive context that shapes the framing process (see Pettenger 2007). This is the purpose of this chapter.

The chapter begins by identifying the dominant frames though which the 'climate change' challenge is communicated to the public. The section then moves into the construction of the scientific message itself through the dominance of certain methodological approaches in researching it. The following section broadens the focus to the transition to a low-carbon society, namely the intersection of science and policy, through critically analysing the processes of how it is studied including simulation modelling. The third section looks at the wider context of the global system and how this is understood in framing the challenges of climate change. The penultimate section returns to the discussion of technology begun in Chap. 1, highlighting the centrality of technology to the dominant framings of the issue of climate change, and identifying the limitations that this imposes. The final section draws conclusions.

Climate Change: Dominant Frames

In studying the framing of the climate change message, scholars of political communication have identified a series of frames through which the message could be tailored to fit the values, perceptions and attitudes of different audiences, ensuring that they can make sense of it. Nisbet lists eight frames, as adapted and simplified in Table 2.1.

He sees frames as 'interpretive storylines that set a specific train of thought in motion, communicating why an issue might be a problem, who or what might be responsible for it, and what should be done about it'. It is 'an unavoidable reality of the communication process, especially as applied to public affairs and policy' (Nisbet 2009: 3). A frame can emphasise an aspect of the issue to make it fit an audience's pre-existing interpretations.

Table 2.1 Typology of climate change frames

Frame	*How it defines the issue*
Social progress	Being in harmony with nature
Economic development	A contribution to growth and jobs
Morality and ethics	Right and wrong; respecting boundaries
Scientific and technical uncertainty	Claiming a lack of consensus and uncertainty about the issue
Frankenstein's monster	Possible catastrophic consequences
Public accountability	Ensuring policy is based on scientific evidence
Alternatives	Seeking to reconcile polarised views
Conflict and strategy	An elite-driven debate

Adapted from Table 2 in Nisbet (2009)

Much has been made therefore by conservative groups of the frame of scientific and technical uncertainty, effectively minimising the threat from climate change, and even its very existence.[1] The Danish political scientist Bjørn Lomborg, and his book *The Skeptical Environmentalist* (Lomborg 2001), is one influential example of this approach. On the other hand, some perceive that environmentalists have tended to counter this with an alarmist message, what Nisbet calls 'Frankenstein's monster', focusing on looming crises and devastation.[2] Former US Vice-President Al Gore seeks to promote an awareness of the considerable risks of climate change though his book *An Inconvenient Truth* (Gore 2006) and the educational work of the Climate Reality Project in which he is actively involved, but this framing is criticised by some on the basis of particular political persuasions, rather than scientific ones. In this situation, it is not surprising that for some sectors of the media the story becomes what is seen as a conflict between experts, the so-called 'conflict and strategy' frame, such as the 'climategate' controversy on the eve of the 2009 Copenhagen climate summit.[3]

Any consideration of the frames within which the issue of climate change has been debated must take note of the active campaign by vested interests and ideological opponents. As part of the landscape of identity and power dynamics in the world today they have sought to lever public opinion. Powerful media organisations such as *The Wall Street Journal*, *The Daily Mail*, *The Telegraph* and *The Sunday Times* have been at the forefront of efforts in the media to discredit the science and action on climate change. As was noted by Ravindranath: 'As the reputation of the IPCC soared, the right-wing climate sceptics also became aggressive in the virulence of their attack on the climate change phenomenon itself' (Ravindranath 2010: 26).[4] Oreskes and Conway (2012) documented parallels between attempts to generate controversy about global warming and earlier public concerns such as tobacco smoking and cancer, acid rain, DDT, and the hole in the ozone layer. They showed that in each case 'keeping the controversy alive' has been attempted after a scientific consensus had been reached. Right-wing think tanks and industry lobbyists funded by fossil fuel interests have sought to misrepresent the normal 'uncertainty' in the finer points of a science, and extrapolate this to a generalised uncertainty on the conclusions, in efforts to spread doubt and confusion among non-scientists and the public.

A frame that has become more dominant since the economic recession of 2008 is to see climate change policy as offering opportunities for job creation and economic growth through investments in renewables, retrofitting homes and developing greener technology such as electric vehicles. This is Nisbet's economic development frame and it has found expression in discussions of a 'green new deal' (NEF 2008). Other more positive social progress frames present the climate change challenge in ways related to lifestyle choices such as diet, recycling or mobility, sometimes labelled green or ethical consumerism (see, for example, Bunyard and Morgan-Grenville 1987). On occasion, a more moral or ethical frame can emerge such as happened in June 2015 on the publication of the encyclical letter of Pope Francis, *Laudato Si*. However, this frame tends to be occasional at best. Nisbet concludes: 'Despite two decades of ever-stronger scientific consensus and record amounts of news coverage, the United States still appears locked in a perpetual divide over climate change, particularly along partisan and ideological lines'. The 'interaction between partisanship and selectively framed media portrayals' results in 'two Americas of climate change perceptions', he writes (Nisbet 2009: 7). While very polarised views find public expression in the US (during the 2016 presidential election campaign, for example), a similar spectrum of perceptions of climate change seems to exist globally.

While it is clear that some perceptions are more in line with what is known about climate change (in its current impacts, the range of potential future impacts and also the potential impacts of policies to address transition), identifying such a spectrum of perceptions helps make visible the range of frames through which the challenge is understood and communicated in the public domain. To some extent, this process of communication tends to rest content with explaining these by reference to tailoring the message to the attitudes and values of audiences. This does not account for the ways we understand the problem itself, prior to the ways it is communicated to audiences, which are themselves framed in particular ways. This is the deeper dimension of framing that needs to be examined and explained. However, the framing of the message by communicators gives us a clue about where to begin and how to proceed. For it makes clear in a rather crude way the link between constructing the message on the one hand and the realities of power politics on the other, in other words, the relationship between science and politics. It is this relationship, and its influence on the way the problem is researched and understood, that needs examining.

The 'science-policy nexus' is an issue that has pointedly cropped up with climate change due to the need for 'objective' information on one hand

and informed policy on the other. A view could be articulated that science and politics are two separate worlds, where scientists see their role as being to provide facts that are as objective as possible, which the policy-makers can then use to fashion policies that can address the problem effectively. This view is based on widely accepted distinctions between facts and values, between the objective world of science and the more subjective and value-laded world of political debate and decision-making. To address this wicked problem, each side knows it needs the other. This is often called a technocratic model of decision-making where the effort is made to base policies on the best scientific evidence and to ensure as little distortion as possible by vested interests. Social constructivism focuses on the ways in which the scientific message itself is socially or discursively constructed, and is influenced by the power realities that structure society. It therefore offers a challenge to the technocratic view, highlighting how even the very best of our scientific evidence comes through processes that are socially constructed. The gravity of climate change as a civilisational issue, and the changes required of our societies and economies, place values, ethics and politics at the heart of the debates. Within these considerations lie issues of identity, values and power-relations, who will benefit and who will lose, and how this influences the science and the message. A question that then arises is how this issue is handled within the science of climate change and the discussion of policy responses, and what are the core issues this has unearthed.

In the Special Report on Emission Scenarios of the IPCC this issue was prominently flagged as it was stated that the definition of 'development,' involves social and cultural dimensions that cannot be resolved by scientific questions, as they become issues of values, preferences and policies (Nakicenovic et al. 2000: 114). Quantitative computer models that explore future socio-economic development in terms of energy, economy, technology and emissions are a key tool in understanding climate change. They allow us to explore the future pathways on which GHG emissions could develop, followed by the related physical impacts of climate change and the monetary costs and benefits of different technologies and policy options. Nielsen and Karlsson (2007: 311) highlighted this science-policy nexus issue in terms of the ways that technological and economic rationalities are implicitly embedded in models. This opens the question of the worldview, values and philosophy underpinning supposedly 'objective' scientific information. This information can reflect specific futures that are profitable or preferable to certain interests or can be used to legitimise results rather than guide policy (see Chap. 5). In Chap. 1 we discussed future global fossil fuel demand suggested by the projections of BP, Shell, Exxon and OPEC. All of these organisations have a vested interest in maintaining market share for their products, so that industry forecasts, projections and scenarios of energy and technology therefore need cautious appraisal. Scenario studies do not preclude consideration of alternative political systems and radical social and cultural change that may lead to these pathways, but these tend not to receive attention in analytical emission scenarios. Modelling tends to be built on assumptions of economic growth, particular technology portfolios and on neoclassical economic assumptions about the ability of markets to deliver particular outcomes and on rational market actors that are selfish and welfare-maximising. But as Deakin et al. (2016: 1) state: 'Beyond textbooks and formal models, markets are not naturally self-adjusting'. Designing future scenarios, forecasts and projections are therefore deeply political exercises that make explicit assumptions about the type of economy that may evolve, the rates of growth and the technologies that are deployed. They make implicit assumptions about power and politics, the underlying philosophy and values of the world as competitive or cooperative, and effectively who is winning and who is losing.

Controversy in valuing the damages of climate change occurred during the process of producing the IPCC's second assessment report. In analysing the costs and benefits of preventing global warming, the value of a human life was estimated by some environmental economists at $1.5 million for people from the richest countries down to $100,000 for those in poorer developing nations (Pearce 1995). These are evidently thorny political and ethical issues and the Indian delegation argued that the calculations were 'absurd and discriminatory' and the then Indian environment minister Kamal Nath wrote to ministerial colleagues around the world asking for these data to be 'purged from the process' (ibid.). While the 'willingness-to-pay' of local populations to avoid loss-of-life currently varies between countries due to differences in income, it presents ethical issues in estimating the statistical value of a human life. It has been overcome by applying a global average of the statistical value of a life to address equity concerns, broadly consistent with government policies towards income redistribution (Markandya and Halsnaes 2001: 483). But the equity controversy of these valuations can never fully dissipate.[5,6] Aggregating such costs are fraught with philosophical difficulties and issues of social justice. It remains the case that the damages from climate change will disproportionately impact poorer regions, while the historical responsibility lies with the wealthier countries who have emitted most of the greenhouse gases since the industrial revolution. As an eminent climate economist Terry Barker pointedly noted in representations to the UK House of Lords, 'I do not accept that the extinction of a species or the extinction of the human race for that matter is subject to a monetary valuation' (Select Committee on Economic Affairs 2005: 84).

There are also analytical issues in fully understanding future damages. Despite updates to the estimates of the future damages of climate change modelled by Integrated Assessment Models (IAMs) there are notable limitations in the inclusion of all relevant climate change impacts and risks (Ackerman and Stanton 2012). The key battleground is known as the Social Cost of Carbon (SCC) which seeks to place an economic value on the damage from each unit of carbon emissions to inform the assessment of policies to reduce emissions. However, according to the IPCC Fourth Assessment Report, 'It is very likely that globally aggregated figures underestimate the damage costs because they cannot include many non-quantifiable impacts' (Pachauri and Reisinger 2007: 69).

The IPCC itself has been a remarkable achievement in cooperative global science. Since its creation in 1988 as a scientific and intergovernmental body under the UN, it has sought to address the science-policy nexus with participatory procedures that include global expert review and government sign-off from all participating countries, seeking to be policy relevant but not policy prescriptive. So where does this leave the social construction of science with respect to climate change? All science functions sceptically and should be rigorously questioned to ensure that the outcomes are valid and robust, and that the place of ethics, values and politics has been understood. However, the greatest controversies in climate change have occurred with respect to vested interests, lobbyists and media which have sought to deny its existence, downplay its impacts or discredit policy to address it (Oreskes and Conway 2012). Efforts to manipulate public and political opinion and to frame the debate in a particular way, may be reprehensible, but these efforts are likely to eventually fail. This would be similar to what happened when the link between tobacco smoking and lung cancer was published (Oreskes and Conway 2012). What is at issue is the time that is being lost through such efforts seeking to delay action.

However, there is also scope to question the philosophical and political underpinnings of research efforts on describing the future of our societies and economies in scenarios and modelling. Further interrogating these is not only a challenge to our understanding of climate change, but to economics in general, which has become somewhat removed from its roots in moral philosophy (Sen 1988). However, it is likely that such a process will hold no succour for deniers and vested interests, it will further reinforce questions about the development models we are applying and who they benefit. If any smoking gun is to be found with climate change it is in the desperate attempts to dilute the issue of climate change from the necessary global political significance it has now rightly attained. But as a framing issue of science communication, the way that climate change and its policies are represented, particularly by the media, is a challenge, and as Box 2.1 illustrates, this may not be as straightforward as it seems. For the relationship between carbon emissions and carbon levels shows that even impeccable scientific data raise issues for perception and understanding of environmental science. The next section considers the dominant ways in which the scientific evidence is translated into policy options.

Box 2.1: Managing the Emissions Message

In March 2016 the International Energy Agency (IEA) reported that, for the second year in a row, energy-related emissions of CO_2 were flat. At the same time, the National Oceanic and Atmospheric Administration (NOAA) in the US reported that 2015 had seen the biggest jump in CO_2 levels ever measured. The IEA's report seems good news, confirming that at long last the world is getting on top of the emissions curve and that policies are beginning to work. But if this is the case, why are CO_2 levels jumping?

As Joe Romm has written, many people are confused about this, including well informed people. He uses the analogy of a bath tub filling up; even if the water flowing in through the tap levels off, the bath tub remains full until its water can be drained off. This helps to illustrate what is happening to the atmosphere: we may be stabilising the amount of additional CO_2 being emitted by human activities but the amounts in the atmosphere remain very high due to the slow drainage effect, in other words the inability of carbon sinks to soak them up. And, in fact, certain atmospheric conditions, such as the El Niño oscillation, cause the sinks (ocean and land) to release more CO_2.

Romm reports that land and oceans are becoming steadily less effective at removing excess CO_2 from the atmosphere. This is because global warming is increasing forest and peatland fires, thereby turning a land sink into a carbon emitter. Ever-worsening droughts and defrosting permafrost have the same effect. So, as Romm puts it, 'we are destroying nature's ability to help us stave off catastrophic climate change'.

The science therefore is telling us a lesson that is the direct opposite of what the good news story seems to imply. Instead of taking a wait-and-see attitude believing that our policies are beginning to work and at least stopping the growth of emissions, we face a race against time to reduce emissions swiftly enough to salvage the ability of land and oceans to act as effective carbon sinks (Romm 2016).

Transitioning to a Low-Carbon Society: Framing Options

The critique just outlined derives from examining the intersection between science and society, known as the 'science-policy nexus'. As is clear from the examples given, this is not a neutral space but is deeply configured by power relations, from the statistical value of a life and assumptions about future technology and development, to the power inequalities that structure today's global order. All scientific endeavour takes place in these configured spaces, as does all human activity. In the case of the wicked problem of climate change, this finds expression in the policy agenda that has emerged as the dominant means to address the realities of global warming and its impacts. Policy-makers have adopted the goal of transitioning to a low-carbon economy/society by 2050 if we have any hope of keeping warming to within 2°C (Herring 2012) while the two principal policy objectives being addressed are mitigation and adaptation, mitigation being the reduction of emissions, and adaptation measures to cope with the damages caused by a changing climate.

The IPCC makes clear that stabilising GHG emissions 'will require large-scale transformations in human societies' and in their 5th assessment report they outline a range of 'transformation pathways' (Clarke et al. 2014: 418). The IPCC bases its analysis of the pathways to a low-carbon society on data from over 1000 new scenario studies produced since the 4th assessment report (published in 2007) collected from integrated modelling research groups. A number of the characteristics of these modelling exercises are identified by the IPCC: firstly, a large majority assume the deployment of carbon dioxide removal (CDR) technologies; secondly, estimates of the aggregate economic costs vary widely; thirdly, most studies are based on 'idealised assumptions that all countries of the world begin mitigation immediately, there is a single global carbon price applied to well-functioning markets, and key technologies are available'; and, fourthly, 'the scenario literature does not systematically explore the full range of uncertainty surrounding development pathways and the possible evolution of key drivers such as population, technology, and resources' (ibid.: 418). The 'large-scale transformations in human societies' required to transition to a low-carbon society, therefore, are themselves highly determined by a strong bias towards technocratic solutions and an inadequate consideration of the political and social dimensions of the transitions required. In other words, the ways in which the goals of social transformation are understood and framed are linked to the socio-economic development scenarios produced to examine climate change and environmental issues. This requires a critical examination of the role of modelling in understanding transition.

Holtz et al. (2015) define a model as 'a simplified, stylised and *formalised* representation of (a part of) reality' that facilitates systematic experiments to identify the outcomes of certain technological and policy options (including pricing and behavioural dimensions). They are explicit, clear and systematic, they allow inferences of dynamics in complex systems and they can be used to test policies or approaches for governance indicating how they might affect situations in future (Holtz et al. 2015: 3–4; emphasis in original). Yet, transition research based on modelling highly complex social and economic dynamics which include variables relating to technologies, infrastructure, institutions, actors, behaviours and values, of necessity 'includes deep uncertainties' in future outcomes. Since social processes cannot be easily captured in models due partly to the fact that the agency of a single or a few actors can influence the process and its outcomes, in the opinion of Holtz et al. models often 'miss the point' because 'their dynamics do not incorporate agency where it would be appropriate'. As a result 'the dangers of relying on model forecasts as accurate predictions are severe' (ibid.: 5–6). It is for this reason that alternative scenarios are employed that begin with qualitative consideration of different future development paths and seek to model these as quantitative results. The analytical issue of considering alternative social and political configurations in the scenarios is allied to the communication issue of representing the full range in plausible future outcomes to policymakers, who may prefer the simplicity of single estimates to the complexity of alternatives. Models are widely perceived by policy makers and the general public as offering levels of scientific certainty but they themselves involve conceptual choices that rest upon certain assumptions and values that are not always made explicit.

Evaluating the contribution of modelling to our understanding of the transition to a low-carbon society needs to pay special attention to the interface between the knowledge produced and the policy-making process. How much modelling is based on the assumption of a 'benevolent central planner' (Box 2.2) making rational choices based on the best available scientific evidence? Yet, as Markard et al. state, the transition to a low-carbon society rests on far more than scientific evidence; it requires 'guidance and governance' since 'a transition is purposeful and intended, and a broad range of actors is expected to work together in a coordinated way'. Political actors, as well as regulatory and institutional support play a vital role but they also work in a context in which the very meaning of what is considered sustainable 'can be subject to interpretation and might change over time' (Markard et al. 2012: 956–957). As a result, transitions can evolve following different kinds of pathways. Therefore, 'there is a pressing need to improve the understanding of the politics and policies of sustainability transitions. At a more conceptual level, issues of power and politics had originally been somewhat neglected'. They ask some pertinent questions of transitions researchers: 'Where (with whom) does power reside in transition processes? How are power and agencies performed in transition processes? Whose voices and narratives remain unheard? Which transitions are legitimate and how can this be assessed?' (ibid.: 962).

Box 2.2: Resorting to 'A Benevolent Central Planner'

In seeking technical advice on developing a low-carbon roadmap for Ireland with the aim of transitioning to a low-carbon, climate resilient and environmentally sustainable economy up to 2050, Ireland's Department of the Environment commissioned the Environmental Research Institute (ERI) at University College Cork, which used a modelling tool known as TIMES, a tool supported by the International Energy Agency (IEA). The model is a 'techno-economic linear optimisation model with the objective of producing a least cost energy system subject to defined constraints' (Deane et al. 2013: 11).

Acknowledging that the challenge of decarbonising the energy system is 'an enormous and expensive one', the ERI finds that TIMES 'produces energy pathways over multiple time slices for a long-term horizon and the solutions of the scenario runs is in terms of technology choice'. It also provides 'indicative results for the carbon price required to achieve certain reductions' which can inform policy design. In the absence of a modelling framework, any analysis of the energy system over the coming decades 'would revert to educated guesswork', state the authors (ibid.: 12).

However, they caution that results of the energy system model 'should not be considered as forecasts for the future'. Instead, they provide 'insights into the impacts of a particular scenario', based on a discrete set of input assumptions in relation to variables such as macroeconomic drivers, fuel prices, resource availability and technology cost. 'These assumptions should not be seen as prescriptive, but rather as a snapshot of potential outcomes that may be realized' (ibid.: 11).

Moreover a number of limitations of the model are identified. These include the macro-economic assumptions on which it is based, its limited capacity to simulate behavioural aspects (basing it on responses to pricing), and incomplete consideration of the extra costs associated with expanding the gas network, shipping ports or electrical transmission costs which are 'considered in a simple manner' (ibid.: 13).

'The modeling perspective taken in this analysis is that of a benevolent central planner: as if there was a single decision-maker taking rational choices surrounding all energy-related issues on technologies and fuels at the lowest cost to the economy and to society. This clearly does not reflect reality, where there are many decision makers and not all decisions are rational, but it does provide very useful guidance into how to achieve CO_2 reductions to 2050 using a least-cost approach' (ibid.: 12).

Yet, the very production of the scientific knowledge on climate change policy can have the result of focusing transition studies on the engineering of mitigation and adaptation, while neglecting issues of power in society and the competing values that inform different social visions that could help incubate competing projects for society. This is evident in the IPCC's analysis of 'transformation pathways' which focus on ways to reduce emissions and to adapt to the impacts of a changing climate. Studies done in both developed and developing regions reflect the same approaches. Ghersi examines a range of studies modelling the transition to a low-carbon society done in Britain, France and Germany, both those commissioned and carried out by public bodies and those by NGOs such as Greenpeace. He identifies a wide range of policy recommendations from these studies addressing technologies, policies and practices for mitigation and for adaptation. Yet, he finds 'a striking gap between the wide array of policy instruments envisioned in the policy-making literature and the scarce modelling expertise on policies and measures beyond carbon pricing'. He finds modelling studies are too isolated from a scientific literature that is grappling with 'the real-life complexities' of transition and from a policy-related literature 'whose diversity echoes the same real-life complexities' (Ghersi 2014: 357).

Surveying the Asian Modeling Exercise (AME), a series of studies modelling the transition in China, India, Japan, South Korea and Nepal, Kainuma et al. illustrate the ways in which hard data on the energy intensity of GDP and the CO_2 intensity of energy apply a range of assumptions such as the large-scale reliance on nuclear energy and a moderate reliance on carbon capture and storage technologies (CCS) as yet to be developed. Furthermore, pathways to low-carbon societies in these countries rest on assumptions about minimising welfare losses, on consumers selecting low-carbon seasonal foods, on environmentally enlightened business and industry, and on a switch to pedestrian and cycle-friendly transport. Yet, they acknowledge that 'it is not an easy task to link the modelling outputs to governmental policies' (Kainuma et al. 2012: S323). Again, treatment of political and social power issues rest on extremely benign assumptions devoid of the realities of contestation and conflict.

Not only is the climate change message presented through framing devices, but so too are the dominant policy options of how we might transition to a low-carbon society. While quantitative modelling makes a contribution to elucidating some dimensions of the available options, particularly in regard to energy, transport, housing and other key sectors, on the key variables of political and social power it rests on idealised assumptions that severely oversimplify the complexities and uncertainties involved. It therefore runs the risk of lulling policy makers and the general public into a complacency about the scale of the challenges facing society and the need for rigorous examination of what type of pathways would be adequate to take us to a low-carbon society by 2050. Consideration of pathways needs to return to the real world in which the transition has to take place.

The Wider Context: Global Political Economy

In outlining how the societal transitions research field understands its subject matter, Holtz et al. adopt a definition taken from Rotmans and Loorbach (2009) who define it as 'a radical, structural change of a societal (sub)system that is the result of a coevolution of economic, cultural, technological, ecological, and institutional developments at different scale levels'. Transitions are taken 'to cover key areas of human activity, including our transport, energy, agrifood, housing, manufacturing, leisure and other systems'. In studying change in these systems, researchers in this field adopt 'a broader perspective than other approaches to sustainable development' and highlight 'the multi-dimensional interactions between industry, technology, markets, policy, culture and civil society'. They recognise that transitions 'are highly complex processes that unfold over time-spans of decades rather than years, and involve "wicked" problems for societies that require a systems approach to policy' (Holtz et al. 2015: 2).

This way of grappling with the complexities of social change begins to situate it within historical trajectories and structured socio-economic and political systems. However, in considering radical structural systemic change as 'the coevolution of economic, cultural, technological, ecological, and institutional developments at different scale levels' a greater focus on identification of the key actors involved through agency and networks of influence is necessary (Hughes and Strachan 2010). While the dynamic of real-life interactions of agency and structure remains contested in the social sciences, transition studies must offer workable and robust conclusions on how society evolves to which 'systems thinking' and hybrid scenario approaches are contributing constructively to transcend the difficulties.

Another starting point for understanding how agency-structure interactions happen is to focus on how paradigms shape them, facilitating some actions and constraining others. Mitchell identifies how UK policy on stimulating the development of sustainable energy technologies rests on a particular paradigm, what she calls 'the Regulatory State Paradigm (RSP)'. This assumes that the market is best placed to select the means to achieve the objectives sought, within a broad regulatory framework set by the state. However, she argues that this approach 'is unlikely to be sufficient given the need to radically redirect the economy in order to respond to the threat of climate change'. She explains:

> There is a danger of ideological 'lock in'. A political paradigm establishes its own institutions and those institutions initiate policies based on the principles of the paradigm—currently reliance on market competition as the main arbiter of value. Those principles and policies promote narrow, short-term, economic considerations which are unlikely to deliver the technical, industrial, institutional and human innovations required. (Mitchell 2010: 1)

She argues for a political paradigm shift if we are to be more able to deal with the climate change challenge. It would involve more government regulation, taking a wider socio-economic view to stimulate the development, deployment, acceptance, take-up and use of relevant technologies and associated infrastructure, understanding innovation from a systems perspective rather than 'the current narrow technological perspective' and the incorporation of qualitative social science perspectives into the policy framework (ibid.: 2).

Box 2.3: Aligning Policies for the Transition

In 2014, the Organisation for Economic Co-operation and Development (OECD) was tasked by Ministers from its member countries to work with the International Energy Agency (IEA), the Nuclear Energy Agency (NEA) and the International Transport Forum (ITF) to better align policies 'for a successful economic transition of all countries to sustainable low-carbon and climate-resilient economies' (OECD 2015: 2). The report, entitled *Aligning Policies for the Transition to a Low-Carbon Economy*, was published as a contribution to preparations for the Paris climate summit in December.

The report identifies a range of policies that risk hindering the transition. For example, in the transport sector there are gasoline or diesel prices that do not reflect the full cost to society, subsidies to company cars, under-valued property taxes, zoning rules that discourage dense building, taxes on property transactions and a lack of local government co-ordination for infrastructure investment. Together these would lead to high levels of CO_2 emissions.

The report's key recommendations centre on scaling up sustainable low-carbon investment and finance, eliminating subsidies and tax expenditures that favour the production and use of fossil fuels, identifying trade barriers that undermine climate objectives, decarbonising electricity through new market arrangements that offer long-term price signals, more energy-efficient and less carbon-intensive mobility, and strengthening incentives for sustainable land use. 'An ambitious climate action plan requires new approaches to policy-making across government,' says the report (ibid.: 4).

The report acknowledges that measures such as carbon pricing and the removal of fossil fuel subsidies 'rarely proceed without resistance from those who stand to lose out economically in the short term'. To address 'outright rejection', it recommends that governments 'align social policies and design compensation schemes in anticipation of the regressive effects of climate policies' (ibid.: 220). Climate pushes policy makers into uncharted territory since it implies the decline of some economic activities and the growth of others. 'Governments must address these issues proactively' (ibid.: 222).

While many of Mitchell's proposals are similar to those of the OECD's recommendations on aligning policies to facilitate the transition to a low-carbon society (see Box 2.3), she makes visible a fundamental dimension of what is constraining the adoption of more adequate policies, which is largely neglected in the voluminous literature on the subject. For she challenges the dominant understanding of the relationship between science and policy, exposing the reality that policy is much more influenced by the prevailing dominant paradigm than by the evidence of science. She describes it as 'a band of iron holding together a certain framework' so that the framework 'constrains certain actions or policies'; until this band of iron is broken, she writes, 'the UK can only do so much and no more in its quest to move to sustainable development' (ibid.: 2). And what holds for energy policy in the UK, holds equally for the range of policies to transition to a low-carbon society in countries throughout the world. Yet, for the OECD proactive action by governments based on scientific evidence (both climate science and policy studies) is sufficient to address the misalignment of policies. Mitchell's analysis takes the issue further since it identifies that the dominant paradigm limits what is possible to do. What we need to identify then is the dominant political paradigm, how this is constraining the actions necessary to transition to a low-carbon society, and what sort of paradigm might better facilitate and stimulate such actions.

In their essence, paradigms structure particular power relations. For example, Mitchell's outline identifies as loci of power the power of privatised energy companies, the nature of state regulation of these companies, the consumption practices of citizens and the nature of technological innovation. It is the interrelationships between these that configure the dominant paradigm. And, as is clear from the example just given, it is not just political power that configures the paradigm, but the ways that political and economic power interact. It is therefore a political economy paradigm. Power in all societies is structured through particular political economy paradigms in which the interrelationships of political and economic power, or state, market and society as it is often put, profoundly shape social outcomes. While these have local variants in each country depending on the nature of the economy, the political system and culture, the social structure and cultural currents, political economy analysts have identified three ideal types focusing on the dominance of state, of the market, or of social class, as shown in Table 2.2 (where TNCs refers to transnational corporations). This shows that each political economy model rests on a range of viewpoints that derive from wider worldviews or ideologies, each of which favour certain actors and actions over others. Looking at climate change through this prism also helps identify how the different ways it is seen are configured by political economy assumptions.

Just like other frames considered in this chapter, political economy can itself be regarded as a frame through which complex issues are understood and which contributes to designing responses that are adequate. Its strength lies in recognising the many ways in which power is structured and diffused throughout society. It therefore operates to complicate the

Table 2.2 Political economy models

Factors	*Statist*	*Market*	*Society*
Ideology	Mercantilist	Liberal	Critical/Marxist
Key actors	States	Firms	Class, social groups
View of TNCs	Beneficial/harmful	Beneficial	Exploitative
View of human nature	Aggressive	Co-operative	Malleable
Behavioural dynamic	State as rational actor	Individual as key actor, not always rational	Struggle between classes, gender
View of market	Needs to be regulated	Positive	Exploitative
View of climate change	Requires global action	Market opportunity	Threat to consumption/ NGO mobilisation

Source: Authors based loosely on Table 1.2 in O'Brien and Williams (2013: 21)

parsimony that informs approaches based on modelling that have been so influential in the design of pathways to a low-carbon society. While many of the frames we have considered operate through focusing on certain aspects while neglecting or marginalising others (treating them as residual or easily resolvable), political economy seeks to integrate as wide a range of issues as possible placing the focus on the whole system and the dynamics that structure it. While much of the national level studies of the low carbon-transition have relied solely on techno-economic modelling exercises, the use of integrated qualitative and quantitative scenarios at the global level has sought to apply systems thinking that reflects this holism (Morita et al. 2001)[7] and overcome the limitations of purely quantitative methods (O'Mahony 2014). There are benefits to enhancing the inclusion of political economy in the analytical and policy approaches to climate change, and in including climate change in political economy literature. This is an endeavour which requires more attention in the debates on how to transition to a low-carbon society. We return to consider political economy models and their contribution to addressing climate change in more detail in Chap. 4.

Technology: What Society Are Our Tools for?

In all this discussion, the role of technology looms large. Indeed, it can be said to be the predominant frame that defines all others in various ways. The dominance of the technological paradigm in economic and political life has already been identified in Chap. 1. Here we focus more closely on how this paradigm defines and limits responses to climate change. In his analysis of technology as a 'hazardous concept', Marx makes clear that the hazards are conceptual, not physical. It is not the dominance of the artefacts of technology that is the primary problem, but the ways of thinking relating to the role of technology in shaping society and social change (Marx 2010).

Marx sees two main problems to be investigated. The first is the ideological, namely the ways that, in the nineteenth century, the emergence of technologies like the steam engine and the telegraph, led to a subtle shift in the Enlightenment view of progress. Instead of technological progress being seen as a *means* to achieve social progress within the frame of a wider republican view of the political and social objectives to be achieved (democracy, social equality), gradually technology came to be seen as the embodiment of progress. As he writes, innovations like the railway 'represented a socially transformative power of such immense scope and promise as to be a virtual embodiment—a perfect icon—of human progress ... a technical means of arriving at social and political goals'. Thus, the machine became synonymous with progress resulting in 'the blurring of the distinction between mechanical means and political ends' (ibid.: 566).

The second problem relates to the substantive changes to socioeconomic systems that happened as a result of the emergence of new technologies. For the railway was not simply a discrete new system of transport but it required 'a new kind of sociotechnological system' that went far beyond the machines themselves. This required:

1. several kinds of ancillary equipment (rolling stock, stations, yards, bridges, tunnels, viaducts, signal systems, and a huge network of tracks);
2. a corporate business organisation with large capital investment;
3. specialised forms of technical knowledge (engineering, telegraphy);
4. a specialised trained workforce with unique skills to keep the system functioning 365 days a year (engineers, firemen, telegraph workers, brakemen, conductors);
5. and institutional changes allowing the system to operate smoothly (regulations standardising track gauges and a national system of time zones).

The scale and capital requirements of this system soon led to the emergence of the anonymous public corporation to replace the private family firm as the dominant institution of capitalist production and exchange, leading to a new kind of professional and scientific management of the economy (ibid.: 567–568).

The emergence of this new socio-technological system also has, as the American economist and sociologist Thorstein Veblen (1857–1929) put it 'become a cultural force of wide-reaching consequences' (quoted in Marx, ibid.: 572), transforming 'the mental habits and, most importantly, the moral and metaphysical assumptions of those who worked with it' (ibid.: 572). There are various aspects to this that find expression in the framing of climate change as analysed in this chapter. The first is what we can call its amorality; as Veblen put it: '[It] gives no insight into questions of good and evil, merit and demerit. ... The machine technology ... can make no use of any of the attributes of worth' (quoted in Marx, ibid.: 572). The second is that it takes on its own logic: technological progress is seen as a good in itself, driven by its own innovations and by the economic need of its owners for profits. The third and most important is that 'it distracts attention from the human—socio-economic and political—relations which largely determine who uses them and for what purposes' (ibid.: 576). The technology itself takes on agency and the machine is seen as the cause of social change rather than the social groups controlling it and using it for certain objectives. Marx's summary of the hazards of technology highlights limitations in the frames through which we understand and address climate change, and the aspects that we neglect or marginalise:

> Technology, as such, makes nothing happen. By now, however, the concept has been endowed with a thing-like autonomy and a seemingly magical power of historical agency. We have made it an all-purpose agent of change. As compared with other means of reaching our social goals, the technological has come to seem the most feasible, practical, and economically viable. It relieves the citizenry of onerous decision-making obligations and intensifies their gathering sense of political impotence. The popular belief in technology as a—if not *the*—primary force shaping the future is matched by our increasing reliance on instrumental standards of judgement, and a corresponding neglect of moral and political standards, in making judgements about the direction of society. (ibid.: 577; emphasis in original)

Box 2.4: President Higgins on Reconciling Ethics, Economics and Ecology

Addressing the mayors of the global 'sister cities' network who met in Dublin in April 2016, President of Ireland Michael D. Higgins urged 'the need for a reconciliation between ethics, economics and ecology ... as an essential dimension of any adequate response to climate change'. While praising technological innovations such as electric cars that can reduce our ecological footprint, he urged the mayors that the concept of 'smart cities' does not become 'mere rhetorical cover for the commercial strategies of, for example, big technology companies'.

The search for innovative technological and scientific solutions must be complemented by 'an awareness of those wider power issues', he said, since such technologies 'are always at risk of being blocked by powerful interests, and in particular those tied to the exploitation of fossil fuels'. Furthermore, cities' responses to climate change must look beyond 'regulations and action plans implemented by urban authorities':

> It must go further to comprise an examination of lifestyles and political decisions that concern all our citizens. Indeed any adequate strategy for tackling climate change at city level is one that should, I believe, be predicated upon the needs of citizens, and in particular the most vulnerable amongst them. We should aspire to an urban civilization and culture of sufficiency as an alternative to contemporary exhortations to insatiable consumption.

In doing this, the President said, we have 'so much to learn from the extraordinary creativity deployed by the people of the shanty towns of India and Egypt, of Peru's *pueblos jóvenes* or of Brazil's *favelas*'. 'Smart cities' are not just about science and new technologies, 'they are about people and livelihoods' (Higgins 2016).

The respected climate change scientist Kevin Anderson has drawn attention to what he regards as 'the optimistic spin' put by the IPCC itself on its analysis of what needs to be done to address climate change. He argues that this is not solely the failure of incisive journalism but 'is also the outcome of repeated and questionable commentary from some experts engaged in the IPCC process' and, indeed, of the conclusions 'from many highly complex integrated assessment models whereby an understanding of prices, markets and human behaviour is brought together with the physics of climate change to generate "policy-relevant" and cost-optimised emission scenarios' (Anderson 2015: 1). He recommends to his scientific colleagues that 'as we massage the assumptions of our analysis to fit within today's political and economic hegemony, so we do society a grave disservice' (ibid.: 3) since the gravity of the situation faced by humanity is constantly minimised. Anderson's analysis suggests a pervasive influence of the limitations of the socio-technological system in allowing technological and economic biases constrain and override wider social and political realities. As highlighted by President Higgins (Box 2.4), these realities need urgent attention if we are to transition to a low-carbon society in a way that is just and that faces the radical social and economic changes required. It is neatly summed up by Pope Francis's conclusion that 'we fail to see the deepest roots of our present failures which have to do with the direction, goals, meaning and social implications of technological and economic growth' (Pope Francis 2015: par 109).

Conclusions

This chapter has examined the frames through which we understand and address the wicked problem of climate change. It has moved from the ways the message is communicated, through the ways it is constructed in scientific research methodologies, and on to the interface between the science and policy-making. The chapter has highlighted an insufficient coverage of issues of power and of social agency in the framing of the problem of climate change that derives from the dominance of a socio-technological frame in today's mainstream approaches towards theorising and understanding processes of social change. The chapter introduced the frame of political economy as an alternative way of understanding and addressing climate change, though one that has received insufficient attention in the mainstream literature and discourse. The analysis in this chapter highlights the need 'to engage in alternative ways of imagining and organising society, the economy and people's relationship with the environment as a cornerstone of sustainability thinking for the 21st century' (Rau et al. 2014: 187). We now turn to examine these alternatives.

Notes

1. That climate change is indeed occurring, that it is driven by human activities, predominantly the burning of fossil fuels, and that it is currently having impacts on human and natural systems, is something that is well accepted in the science (IPCC 2014). It is only in some media and related public opinion that doubt or confusion remains. A position of denial is not supported by the empirical evidence and is an example of where science communication has been twisted by vested interests (Oreskes and Conway 2012).
2. It is known in the science that there is a risk of future breakdown in social, environmental and economic systems from climate change as the impacts multiply through the twenty-first century. While it is indeed quite possible to avoid many of the most serious risks through effective action for transition and adaptation considerable risks will still remain, and require inclusion in problem-framing. On the other hand, a scenario of benign impacts of climate change, even with effective transition to a low-carbon future, is very unlikely. Even at 2°C change, unique and threatened systems, extreme weather events and distributional impacts are subject to high risk, with moderate risk for 'global aggregate impacts' and large-scale singular events (IPCC 2014).
3. The 'climategate' controversy which arose from the hacking of emails from climate scientists, was seized upon by deniers with the intention of showing manipulation of climate data. It has been successively debunked as a non-issue by eight different committees of investigation. Finding no evidence of fraud or scientific misconduct, the reports called on scientists to avoid any such future allegations by opening up access to supporting data, processing methods and software.
4. Another such controversy over the inclusion of poor evidence to support a claim of Himalayan glaciers melting by 2035 in the IPCC Fourth Assessment Report (AR4) has been acknowledged as an oversight during the review process (Ravindranath 2010). However, it did not affect any of the core conclusions of the report.
5. Other ways of valuation include a human capital approach, which values the loss of income and multiplies it by the change in risk, or a 'life years lost' approach based on the willingness-to-pay for life years that could be lost as a result of changes in the survival probabilities an individual faces (Markandya and Halsnaes 2001: 483).
6. From a global perspective this has appeal, but national perspectives and opportunities should be addressed in other ways according to Markandya and Halsnaes (2001: 483).
7. Human agency over the future is a concept advanced in 'futures thinking' and scenarios. De Jouvenel (2000: 38) describes the future as a realm of freedom, power and will. Although power may be unequal, 'all the actors have some individual power enabling them to act'.

References

Ackerman, F., and E.A. Stanton. 2012. Climate Risks and Climate Prices: Revisiting the Social Cost of Carbon. *Economics* 6 (2012-10): 1–25.

Anderson, Kevin. 2015. On the Duality of Climate Scientists. *Nature Geoscience*, October 2015. Accessed 21 April 2016. http://www.nature.com/ngeo/journal/vaop/ncurrent/full/ngeo2559.html

Berger, Peter L., and Thomas Luckmann. 1991. *The Social Construction of Reality: A Treatise in the Sociology of Knowledge*. London: Penguin.

Bunyard, Peter, and Fern Morgan-Grenville. 1987. *The Green Alternative: Guide to Good Living*. London: Methuen.

Clarke, L., K. Jiang, K. Akimoto, M. Babiker, G. Blanford, K. Fisher-Vanden, J.-C. Hourcade, et al. 2014. Assessing 'Climate Change 2007: Mitigation. Contribution of Working Group III'. *Climate Change and Sustainable Development* 3 (S1): S19–S40.

De Jouvenel, H. 2000. A Brief Methodological Guide to Scenario Building. *Technological Forecasting and Social Change* 65: 37–48.

Deakin, S., S. Stern, R. Kaplinsky, F. Muniesa, M. Nabli, M. O'Neill, H. Ortiz, K. Sahlin, A. Schwittay, and L. Talbot. 2016. Chapter 6: Markets, Finance and Corporations: Does Capitalism Have a Future? In Draft report for comment of the International Panel on Social Progress. Accessed 23 February 2017. https://comment.ipsp.org/chapter/chapter-6-markets-finance-and-corporations-does-capitalism-have-future

Deane, Paul, John Curtis, Alessandro Chiodi, Maurizio Gargiulo, Fionn Rogan, Denis Dineen, James Glynn, John FitzGerald, and Brian Ó Gallachóir. 2013. *Technical Support on Developing Low Carbon Sector Roadmaps for Ireland*. Cork: ERI and Dublin: ESRI.

Ghersi, Frédéric. 2014. Low-Carbon Policy Making vs. Low-Carbon Policy Modelling: State-of-the-Art and Challenges. *Environmental Modeling and Assessment* 19: 345–360.

Gore, Al. 2006. *An Inconvenient Truth*. London: Bloombury.

Herring, Horace, ed. 2012. *Living in a Low-Carbon Society in 2050*. Basingstoke: Palgrave Macmillan.

Higgins, Michael D. 2016. Speech by President Michael D. Higgins at the Independence and Interdependence Summit "The Role of Cities in Relation to Climate Change", Croke Park, Dublin, 22 April 2016.

Holtz, Georg, Floortje Alkemade, Fjalar de Haan, Jonathan Köhler, Evelina Trutnevyte, Tobias Luthe, Johannes Halbe, et al. 2015. Prospects of Modeling Societal Transitions: Position Paper of an Emerging Community. *Environmental Innovation and Societal Transitions*. doi:10.1016/j.eist.2015.05.006.

Hughes, N., and N. Strachan. 2010. Methodological Review of UK and International Low Carbon Scenarios. *Energy Policy* 38: 6056–6065.

IPCC. 2014. *Climate Change 2014: Synthesis Report.* Contribution of Working Groups I, II and III to the Fifth Assessment Report of the Intergovernmental Panel on Climate Change [Core Writing Team, R.K. Pachauri and L.A. Meyer (eds)]. Geneva: IPCC.

Kainuma, Mikiko, Priyadarshi R. Shukla, and Kejun Jiang. 2012. Framing and Modeling of a Low Carbon Society: An Overview. *Energy Economics* 34: S316–S324.

Lomborg, Bjørn. 2001. *The Skeptical Environmentalist.* Cambridge: Cambridge University Press.

Markandya, A., and K. Halsnaes. 2001. Costing Methodologies. In *Climate Change 2001: Mitigation*, ed. B. Metz et al., 451–495. Cambridge: Cambridge University Press.

Markard, Jochen, Rob Raven, and Bernhard Truffer. 2012. Sustainability Transitions: An Emerging Field of Research and Its Prospects. *Research Policy* 41: 955–967.

Marx, Leo. 2010. Technology: The Emergence of a Hazardous Concept. *Technology and Culture* 51 (3): 561–577.

McGrew, Anthony. 2011. The Logics of Economic Globalization. In *Global Political Economy*, ed. John Ravenhill, 275–311. Oxford: Oxford University Press.

Mitchell, Catherine. 2010. *The Political Economy of Sustainable Energy.* Basingstoke: Palgrave Macmillan.

Morita, T., J. Robinson, A. Adegbulugbe, J. Alcamo, D. Herbert, E.L. La Rovere, N. Nakicenovic, et al. 2001. Greenhouse Gas Emission Mitigation Scenarios and Implications. In *Climate Change 2001: Mitigation, Contribution of Working Group III to the Third Assessment Report of the Intergovernmental Panel on Climate Change*, ed. B. Metz, O. Davidson, R. Swart, and J. Pan. Cambridge: Cambridge University Press.

Nakicenovic, N., J. Alcamo, G. Davis, B. de Vries, J. Fenham, S. Gaffin, K. Gregory, et al. 2000. *Special Report on Emissions Scenarios.* Working Group III, Intergovernmental Panel on Climate Change (IPCC). Cambridge: Cambridge University Press.

NEF. 2008. *A Green New Deal.* London: New Economics Foundation.

Nielsen, S.K., and K. Karlsson. 2007. Energy Scenarios: A Review of Methods, Uses and Suggestions for Improvement. *International Journal of Global Energy* 27 (3): 302–322.

Nisbet, Matthew C. 2009. Communicating Climate Changer: Why Frames Matter for Public Engagement. *Environment: Science and Policy for Sustainable Development*, March–April 2009. Accessed 31 March 2016. http://www.environmentmagazine.org/Archives/Back%20Issues/March-April%202009/Nisbet-full.html

O'Brien, Robert, and Marc Williams. 2013. *Global Political Economy.* Basingstoke: Palgrave Macmillan.

O'Mahony, T. 2014. Integrated Scenarios for Energy: A Methodology for the Short Term. *Futures* 55: 41–57.

OECD. 2015. *Aligning Policies for the Transition to a Low-Carbon Economy*. Paris: OECD.

Oreskes, Naomi, and Erik M. Conway. 2012. *Merchants of Doubt*. London: Bloomsbury.

Pachauri, R.K., and A. Reisinger. 2007. *Climate Change 2007—Synthesis Report*. Geneva: Intergovernmental Panel on Climate Change (IPCC).

Pearce, F. 1995. Global Row over Value of Human Life. *New Scientist*, August 19, 7. Accessed 13 March 2017. https://www.newscientist.com/article/mg14719910-900-global-row-over-value-of-human-life/

Pettenger, Mary E., ed. 2007. *The Social Construction of Climate Change: Power, Knowledge, Norms, Discourses*, 275–311. London: Routledge.

Pope Francis. 2015. *Laudato Si: On Care for Our Common Home*. Vatican City: Vatican Press.

Rau, Henrike, Anna R. Davies, and Frances Fahy. 2014. Conclusion: Moving On—Promising Pathways to More Sustainable Futures. In *Challenging Consumption: Pathways to a More Sustainable Future*, ed. Anna R. Davies, Frances Fahy, and Henrike Rau, 187–205. London: Routledge.

Ravindranath, N.H. 2010. IPCC: Accomplishments, Controversies and Challenges. *Current Science* 99: 26–35.

Romm, Joe. 2016. How Can Global CO_2 Levels Soar When Emissions Are Flat? *Climate Progress*, March 21. Accessed 24 March 2016. http://thinkprogress.org/climate/2016/03/21/3761903/co2-levels-soar-emissions-flat/?platform=hootsuite

Rotmans, J., and D. Loorbach. 2009. Complexity and Transition Management. *Journal of Industrial Ecology* 13 (2): 184–196.

Select Committee on Economic Affairs. 2005. The Economics of Climate Change: 2nd Report of Session 2005–06, Volume II: Evidence. In *Great Britain: Parliament: House of Lords: Select Committee on Economic Affairs*. London: The Stationery Office.

Sen, A. 1988. The Concept of Development. In *Handbook of Development Economics*, ed. H.B. Chenery and T.N. Srinivasan, vol. 1, 9–26. Amsterdam: North Holland.

CHAPTER 3

Addressing the Problem: Understanding Low-Carbon Transition with the Social Sciences

Introduction

'It's the economy, stupid' was a refrain that originated in Bill Clinton's US presidential campaign of 1992 and became synonymous with the idea that the economy tends to take precedence over other interests. The economy may continually rise to prominence in everyday politics, but in the low-carbon transition and in delivering human wellbeing, the place of economy and that of technology could be described as being more like supporting actors. In how we conceive of, and address the challenge and opportunity of delivering a sustainable low-carbon world, the importance of society, as social, cultural and governance factors, and the environment as our life-support system, require much more prominent roles. This chapter offers both theory and evidence as to why these considerations are important, and how we can begin to include them in our thinking.

The ways we conceive, analyse and create policy were related to 'frames' in Chap. 2. The dominant techno-economic paradigm offers insights into the important technological and economic aspects of transition, but leaves gaps in how the overall problem of mitigating GHG emissions is understood. This chapter takes a systemic perspective on transition as first and foremost a development problem, or 'climate change through a sustainable development lens' as described by Sathaye et al. (2007: 696). It emphasises the contribution of systems thinking and the social sciences,

P. Kirby, T. O'Mahony, *The Political Economy of the Low-Carbon Transition*, International Political Economy Series,
DOI 10.1007/978-3-319-62554-6_3

and political economy in particular, in understanding the challenges and the opportunities. The interaction of the driving forces of emissions leads to phenomena such as 'inertia' and 'carbon lock-in' which are discussed in the next section. The Kaya identity (Kaya 1990) is discussed in the section 'Reducing Emissions the Kaya Way' as a useful tool to understand trends and driving forces of emissions, but one which may ignore the issues of 'development' when broadly construed. The holistic concept of 'development paths' is discussed in the section 'Development Pathways' to describe both the development vision of a nation, and how this must be linked to an integrated conception of the driving forces of emissions, including social, cultural and governance factors. As a particular type of development path with many benefits, the section 'Sustainable Development Pathways and Transition' details what '*sustainable development pathways*' may consist of. The section 'Thinking About Mitigation and Transition Through the Energy Hierarchy' gives a specific example of the implementation of a sustainable development pathway in relation to energy, through an energy hierarchy for development and interlinked energy planning. The section 'From Socio-technical Transitions to Sustainable Development' describes the role of the social sciences and political discourse in transition planning through the importance of ethics, values and policies. The section 'Ethics, Values and Policies' discusses socio-technical transitions, a prominent approach to understanding transition, but one which requires augmentation with that of wider change in society. The section 'Conclusion' synthesises the discussion.

Inertia and Carbon Lock-In

The argument of engineers and technologists tends to focus on technical emission reductions, while that of economists has tended to focus on economic potentials and measures. Rosa and Dietz (2012: 584) observed that the social science literature on the drivers of greenhouse gas emissions is fragmented across disciplines, with economists ignoring the work of sociologists, sociologists rarely citing political scientists and so on, with much to be gained by a more cross-disciplinary dialogue. The early years of increased research on climate change and its mitigation pointed to the limitations of deterministic and single disciplinary frames for analysing the problem (Fisher et al. 2007: 175). The systems involved, particularly the human socioeconomic system, are complex, non-linear and multi-dimensional. Understanding these systems cannot rely on technological or economic perspectives alone which do not give the full picture. From the science of climate change to the mitigation of emissions the frame broadened over the years, something that was well illustrated in the IPCC *Special Report on Emission Scenarios* (Nakicenovic et al. 2000) and other global environmental scenario exercises. As the understanding of the processes of change in human systems improved, two important concepts emerged from a more 'systems thinking' approach to energy and other human sources of emissions.[1] These are 'inertia' and the related idea of 'carbon lock-in'.

Inertia means a delay, slowness or resistance in the response of climate, biological, or human systems to factors that alter their rate of change. In the Third Assessment Report (TAR) the IPCC described inertia in human systems as depending on the interaction between social and economic structures and values, institutions, technologies and established infrastructure. The delay in responding arises in the process from awareness of the problem to the implementation of solutions but, on a positive note, it can be influenced by policies and choice (Metz et al. 2001). As human systems change over time a 'path dependency', or lock-in can occur to alternatively higher or lower emissions trajectories with a middle path unlikely (Halsnæs et al. 2007: 150). There is now much concern about the higher emissions trajectories of industrialised countries, and the copying of this pattern in developed countries, as they seek to grow and develop out of poverty to a higher standard of living and improved wellbeing. Regardless of the implementation of low-carbon energy technology,[2] these patterns put the goal of stabilising greenhouse gas emissions and avoiding dangerous climate change at risk. The patterns that emerge include more direct physical drivers such as infrastructure and spatial patterns of development (IPCC 2014: 18). Such patterns depend on the further development of fossil-based energy systems and dispersed low-density human settlement. These increase the costs of mitigating emissions and are difficult or irreversible to change. Nevertheless, a more fundamental challenge exists in the social and cultural dimensions and how they influence governance and policies. Value systems, worldviews and social and cultural norms influence the emergence of the higher emissions lifestyles evident in industrialised countries, and the technologies and energy system that support them. They also provide the context in which decisions are made at all levels: from the daily decisions of citizens and consumers, industry decisions about how to go about business, to public institutions that decide policy and the development choices of a nation. Social, cultural and political factors are continually at play. Where there is cultural and institutional lock-in this can favour more emissions-intensive development patterns in general and more emissions-intensive technology in particular.

It was Gregory C. Unruh in his doctoral thesis at Tufts University who initially fleshed out the concept of 'carbon lock-in', suggesting that escape conditions are unlikely to be generated internally (Unruh 2002). Unruh described the 'techno-institutional complex' of carbon lock-in as consistent market and policy failures generated by the combined interactions of technological systems and their governing institutions. The spread of carbon-saving technologies is resisted and the lock-in extends to social 'institutions, customs and preferences'. While social, cultural and institutional factors may be found influencing not just energy supply and demand and technological issues as described by Unruh, they also influence the very nature of the economy, society, technology and governance, as wider development issues in each country. It is these wider development issues that come into focus in development pathways. One of the key conclusions from this recognition of the importance of inertia and lock-in is that current decisions have long-term consequences. They can embed a long-term development path that limits or prevents future emissions reductions, something further explained in Box 3.1.

Box 3.1: Path Dependence and Lock-Ins

In the IPCC Fifth Assessment Report Fleurbaey et al. (2014: 312) discuss path dependence as a tendency for past decisions and events to self-reinforce patterns, diminishing and possibly excluding the prospects for alternatives to emerge. Path dependency is therefore important in understanding the transition between development pathways. Fleurbaey et al. use the example of developing inter-city highways which make further extension of the road network more likely, and further extension of rail networks less cost-effective, by drawing out traffic and investment, and diminishing the prospects for alternative transportation investments. However, even if policy seeks to successfully control the spatial settlement pattern from becoming dispersed, a phenomenon associated with development of the road network, there are more fundamental systemic consequences. The aspirations of citizens become more associated with private and motorised transport and public policy associates mobility and progress with road-building and the automobile. This cascade effect from the initial decision shows that such choices have effects that are beyond the limits of the original project-level and become self-reinforcing in higher emissions trajectories. This also illustrates why mitigation cannot be seen as an 'environmental' challenge, or one of energy and mitigation. It requires integrated decision-making across government that mainstreams climate considerations, and doesn't simply attempt 'end-of-pipe' technological solutions, but engages with transformation as a development challenge.

Reducing Emissions the Kaya Way

The Kaya identity (Kaya 1990) has been an important quantitative tool for analysis of carbon emissions and greenhouse gases in the climate change debate. It has underpinned such analyses as the IPCC Special Report on Emission Scenarios (Nakicenovic et al. 2000) through its utility in simplifying the driving forces of emissions. Based on the IPAT identity[3] of Ehrlich and Holdren (1971) that was designed to describe the impact of human activity on the environment, the Kaya identity has been important in understanding socioeconomic drivers of emissions both historically and in future climate change. The Kaya identity can be represented mathematically as follows:

$$\text{Total emissions} = \text{Population} \times (\text{GDP} / \text{population}) \times \\ (\text{energy} / \text{GDP}) \times (\text{emissions} / \text{energy})$$

or

$$\text{Total emissions} = \text{Population} \times \text{affluence per capita} \\ \times \text{energy used per unit affluence} \times \text{emissions per unit energy used}$$

The results of the Kaya decomposition of driving forces of global CO_2 emissions can be seen in Fig. 3.1 taken from Rogner et al. (2007: 108) with both historical and future projections to 2030. In determining which driving forces to address in reducing emissions, the Kaya identity focuses

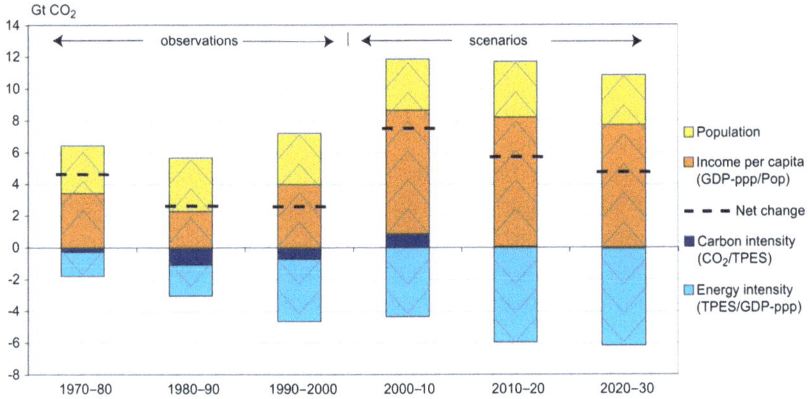

Fig. 3.1 Decomposition of global energy-related CO_2 emission changes at the global scale historically and in the future. Source: Figure 1.6 from *Climate Change 2007: Mitigation of Climate Change. Working Group III Contribution to the Fourth Assessment Report of the Intergovernmental Panel on Climate Change* [B. Metz, O.R. Davidson, P.R. Bosch, R. Dave, L.A. Meyer (eds.)]. Cambridge University Press, Cambridge, UK and New York, NY, USA

attention away from population growth, which is more limited than the other factors in driving global emissions. It is also very difficult to control population outside of supporting women's access to education and literacy, birth control and other socio-cultural factors such as labour force participation. There are notable ethical concerns from past population control policies such as China's one-child policy and forced sterilisations in India. The affluence factor of GDP per capita also tends to get short shrift. Economic growth tends to be a central political goal in almost all countries. While recognising this political reality, the evidence and theory around economic growth is much more circumspect than political discussions would have us believe. The debate on whether it is necessary or desirable to continue economic growth is far from complete (Jackson 2009). This is particularly the case in wealthier countries where the association between income growth and wellbeing is dubious at best (Fleurbaey et al. 2014: 310) and negative at worst (Bartolini 2014). That current forms of growth have negative social and environmental outcomes is supported by much evidence. The theoretical discussions in development studies recognise that the phenomenon of economic growth is a means and not an end of development (Anand and Sen 2000).

Refining the economic development model requires a critique of such outcomes: what kind of growth is favoured, to what level, who does it benefit, and can this be balanced with wider sustainability considerations? The idea of the 'green economy' can be a useful tool to move towards sustainable development, but this depends on how it is applied. If it leads to a focus on eco-efficiency and technological change, rather than addressing the fundamentals of development, it can reinforce problems of inequality and will not lead to the lower emissions development path that is required. The Nobel Laureate development economist Amartya Sen, and the field of development studies in general, consistently criticise the primacy of economic growth. There is a need to avoid the green economy becoming a 'greenwash,' which would allow a continuation of our current skewed development. It will require robust critique of how it serves the development vision in each country, and how this vision supports the wellbeing of people (see Chap. 8).

Shifting the focus, in the last factor of the Kaya identity, carbon per unit energy will show the reduction in carbon emissions through low-carbon energy technology in the form of renewables and nuclear. It also shows decarbonisation of energy supply or 'carbon efficiency' through shifting to fossil fuels of lower carbon intensity such as substituting gas for coal. These purely technological factors are important but not the sole considerations of future transition. It is in the economic energy intensity factor (energy/GDP) of the Kaya identity that many underlying factors of development are hidden. This includes not just technological factors that influence changes in the energy intensity of the economy (through technical efficiency in the production and use of energy), but crucially development factors that influence the energy intensity of the overall society and economy. This is a reflection of how we live, grow and develop. It is these more complex factors that emerge from transdisciplinary perspectives on current trends in emissions and driving forces, and scenario or 'systems thinking' perspectives on future emissions outcomes. Given the factors that are hidden by the Kaya identity, its usefulness in understanding mitigation and transition can be limited. Where the social, cultural and political drivers of emissions are insufficiently characterised, the Kaya identity as currently conceived should be used with a health warning (O'Mahony and Dufour 2015a: 68). This fate has already beset the more simplistic approach of the Environmental Kuznets Curve, which while useful in understanding the trend in some environmental burdens, has provided little insight when it comes to carbon emissions (Stern 2004). The focus on income and economic energy intensity of an entire country in Kaya can be so broad as to become meaningless, if causation is not discussed from both a technological and a development perspective. Technological and development issues are mixed in national economic energy intensity to such an extent that attribution of causation in the drivers of change is difficult (Rosa and Dietz 2012). It is in the idea of 'development paths' that a holistic conception is attempted.

Development Pathways

One of the major conclusions of the IPCC Third Assessment Report (Banuri et al. 2001), the Fourth Assessment Report (Sathaye et al. 2007) and long-term climate scenarios (Metz et al. 2002; Nakicenovic et al. 2000; Swart et al. 2003) was that climate policy alone will not solve the problem but making development more sustainable by changing development paths can make a major contribution. As noted by Halsnæs et al. (2011) shifting from a high- to a low-emissions development pathway could potentially be as important for climate change mitigation as implementing 'climate' policies. Through climate mitigation alone, it will be extremely difficult and expensive to achieve low stabilisation targets (450 ppm CO_2)[4] from scenarios that embody high-emission development paths (Sathaye et al. 2007: 696). A continuing difficulty with this realisation of the importance of the development path is that while it has been known for some time that development paths are crucial to transition, the idea has failed to make notable headway in either policy or analysis beyond the more obvious techno-economic factors.[5] Techno-economic factors are necessary but not sufficient, and understanding what characterises the development path is now critical to transition in all countries.

The rate at which wealthier countries must reduce emissions is a central framing condition in understanding global transition. While most transition envisages a more technically and politically feasible GHG reduction in the order of −3% per annum, studies such as Calverley et al. (2009) suggest that reductions of the order of −9% per annum may be required to be consistent with the science. Within this context, transition would move from incremental change to a gargantuan task. Even assuming emissions reductions of −3% per annum, transition is an urgent priority when it is recognised that short-term decisions have long-term consequences, and that the development pathway can lock-in either higher or lower emissions trajectories. Recognising the difficulties with lock-in is particularly important in developing countries, with gaps in infrastructure

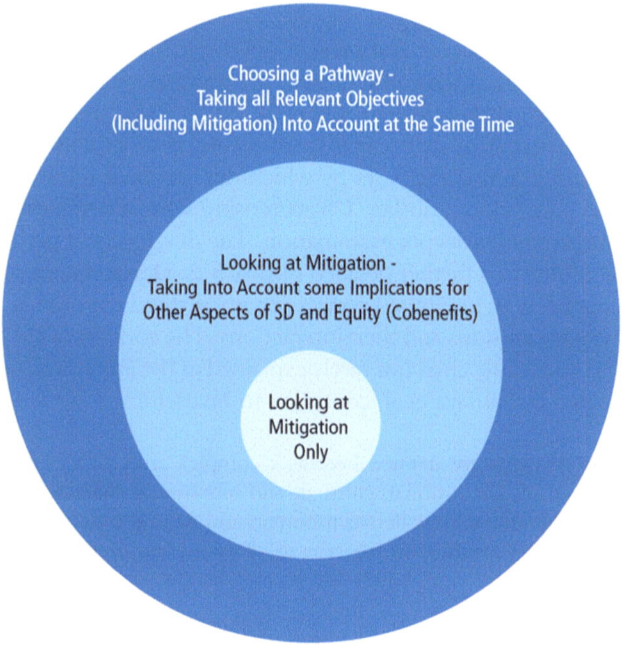

Fig. 3.2 Three frameworks for thinking about mitigation. Source: Figure 4.1 from IPCC 2014: *Climate Change 2014: Mitigation of Climate Change. Contribution of Working Group III to the Fifth Assessment Report of the Intergovernmental Panel on Climate Change* [O. Edenhofer, R. Pichs-Madruga, Y. Sokona, E. Farahani, S. Kadner, K. Seyboth, A. Adler, I. Baum, S. Brunner, P. Eickemeier, B. Kriemann, J. Savolainen, S. Schloemer, C. von Stechow, T. Zwickel and J.C. Minx (eds.)]. Cambridge University Press, Cambridge, UK and New York, NY, USA

development remaining to be filled and high-emissions lifestyles of more developed countries largely absent, lock-in can be avoided. Figure 3.2 identifies three ways of thinking about mitigation, including the earlier more narrow approach of 'mitigation policy'[6] and the more recent approach to consider co-benefits[7] and synergies.[8] The holistic integrated approach addresses not only the previous categories but the overall development path and how they fit into this wider picture.

The understanding of development paths can be described in two main branches: backward-looking and forward-looking. The 'backward-looking' body of work describes past and present development trajectories and their determinants, which includes the growth literature as well as a large part of the development literature (Fleurbaey et al. 2014: 311). 'Forward-looking' studies include scenarios that construct plausible development pathways for the future, and examine the ways by which development might be steered towards one pathway or another. Characterising what a development pathway actually is requires deeper examination. The development pathway as an intellectual tradition is further detailed in Box 3.2. Development paths have been articulated as an integrated concept to understand the driving forces of greenhouse gas emissions and their interactions. The concept of the development path is frequently cited but rarely explained in the literature. The definition articulated by Sathaye et al. continues to be useful:

> Development paths are defined here as a complex array of technological, economic, social, institutional, cultural, and biophysical characteristics that determines the interactions between human and natural systems, including consumption and production patterns in all countries, over time at a particular scale. (Sathaye et al. 2007: 696)

Barker et al. (2007a: 33) described development paths as evolving from economic and social transactions under the influence of a broad range of policy areas from the wider issues of taxes and regulation to the more specific issue of energy efficiency.[9] Another definition was offered by Edenhofer et al. (2014: 50) where a society's development pathway—with its particular socioeconomic, institutional, political, cultural and technological features—enables and constrains the prospects for mitigation. Development paths can be simplified to a first order as the energy, production and consumption systems and the technological choices and political economy factors that influence these. More fundamentally, social, cultural and institutional factors determine what type of development is pursued, what type of economy, society and environment this entails and the role for governance. At a minimum, the development path must be fully integrated across government policy from economic development, energy policy, tax and incentives, spatial planning and research and development, to environmental protection, sustainable development, industrial development, transport, agriculture, food and even related areas such as health and social policy. The failures of national mitigation strategies and indeed sustainable development strategies to adequately integrate their functions with other cross-sectoral strategies have been noted by Casado-Asensio and Steurer (2016: 95). That the relationships between policies are either 'competitive rather than complementary' or 'weak overall' is a clear call for the use of development paths to integrate sectoral policies with the necessary wider vision. The alternative is a continuation of weak climate policy that is more likely to fail to deliver the low-carbon transition, increase costs and conflicts, and reduce the opportunities that change opens up. The results of national mitigation strategies are dubious and have failed to develop synergies or manage trade-offs (Casado-Asensio and Steurer 2016: 100). This is seeking a lose-lose rather than a win-win situation and needs much greater policy attention, to choose national development pathways that contribute to human wellbeing and sustainability.

The development path also includes issues of equity within and across generations, environmental sustainability and nature conservation. It is in each specific context that this must be defined incorporating the 'national way of doing things' (Sathaye et al. 2007: 709). The human factors involved are insufficiently explored in the literature, with frequent generalisations about the importance of social, cultural and institutional factors (Rosa and Dietz 2012), but little substantive discussion of what this means in practice. As a result, mitigation and transformation continue to be dominated by the techno-economic perspective at the expense of a deeper discussion. This is an 'end-of-pipe' approach to the challenge (O'Mahony and Dufour 2015b: 418). The challenge requires framing as one of 'sustainable development' rather than only as climate mitigation, and a recognition that the driving forces of emissions are linked to the underlying development path (Sathaye et al. 2007: 696). Sathaye et al. noted that there is much evidence that making development more sustainable can make a significant contribution to climate goals. As noted by Sathaye et al. a key finding of the IPCC Third Assessment Report (Morita and Robinson 2001) was that low-emission baseline scenarios, may go a long way towards achieving low stabilisation levels even before climate policy is included in the scenario (see Box 3.2).

Box 3.2: Development Pathway as Intellectual Tradition
The concept of a national development pathway is holistic according to Fleurbaey et al. (2014: 311). It is broader than the development trajectory of a particular sector, or of a particular group within a society. A wide range of economic, social, and environmental indicators are necessary to describe a development pathway in quantitative terms, but not all characteristics will be amenable to quantitative representation. This highlights the importance of the inclusion of qualitative inquiry and insights from the social sciences. But as noted by Fleurbaey et al. (2014: 311) 'a "pathway" is not a random collection of indicators. It has an internal narrative and causal consistency that can be captured by the determinants of the interactions between human and natural systems.' The underlying empirical assumption is that in an observed development trajectory, development can be explained as identifiable drivers by various economic, social, and environmental indicators. The concept of development pathway is thus rooted in the dominant intellectual tradition in which history has some degree of intelligibility, as opposed to a chaotic set of unintelligible events in the tradition of Schopenhauer (1966).

Sustainable Development Pathways and Transition

If long-term transition is therefore intimately linked to sustainable development paths, then the sustainability dimensions of national development paths require consideration. The social dimension is strong in addition to the environmental and economic, although it has tended to receive little attention (Boström 2012: 3). For a development path to be sustainable over a long period, wealth, resources and opportunity must be shared so that all citizens have access to minimum standards of security, human rights and social benefits, such as food, health, education, shelter and opportunity for self-development (Reed 1996). There are ethical but also pragmatic reasons for addressing inequality, social justice and environmental sustainability, even from a purely self-interested individual perspective. Growth, security, wellbeing, environmental quality, social capital and a host of other societal indications that could be described as necessary for individual human wellbeing are all dependent on balanced development. Such discussions are widely accepted in the field of development studies, recognising that human wellbeing is multi-dimensional (Stiglitz et al. 2009; McGillivray 2007), and that there are fundamental links between human development and sustainable development (O'Mahony and Dufour 2015b). It is widely known that income inequality has a marked negative impact on individual subjective wellbeing (Fleurbaey et al. 2014: 311). A key point is that the balancing of the three pillars of sustainable development, social, environmental and economic, is not solely about balancing outcomes but is about balancing the processes. While mitigation tends to focus on the techno-economics of technology and cost in reducing emissions, it is the social, cultural and institutional drivers which require far greater attention than has been achieved to date. Chapter 4 of the executive summary of Working Group III of the IPCC Fifth Assessment Report on 'Sustainable Development and Equity' outlines the overarching context of linking long-term transition to sustainable development pathways:

> Governing a transition toward an effective climate response and sustainable development pathway is a challenge involving rethinking our relation to nature, accounting for multiple generations and interests (including those based on endowments in natural resources), overlapping environmental issues, among actors with widely unequal capacities, resources, and political power, and divergent conceptions of justice (high confidence). Key debated issues include articulating top-down and bottom-up approaches, engaging participation of diverse countries and actors, creating procedurally equitable forms of decentralization and combining market mechanisms with government action, all in a particular political economic context (robust evidence, high agreement). (Fleurbaey et al. 2014: 287)

Nonetheless, a useful starting point to consider in choosing a sustainable development path can be related to energy consumption and GHG emissions. These include: mobility and spatial pattern, agriculture and food, income and inequality, consumption and wellbeing, the structure of the economy and trade, and land use and biodiversity. Such a list requires further definition but should not be set in stone, guidance is required but definition in each national context is also necessary. In development issues, the approach should not be too prescriptive until it is defined in a specific social and cultural context as was noted by Sen in framing the capability approach to human development (Sen 1985, 1992). Fundamentally a national sustainable development path is a vision of the balancing of a country's future development of society, environment and economy.

The process of defining sustainable development pathways not only requires more research input but, as a development issue, it requires definition in each national context by participatory processes in pathway planning. It requires active democratic participation in the balancing of the public good against inevitable vested interests and assisting those interests in adapting (IPCC 2012). While the approach to a sustainable development path cannot be too prescriptive in general theories, at the national policy level it must be clearly defined through implementation, monitoring, management and corrective action measures. If a sustainable development pathway is indeed to be effective, and not just rhetorical, it requires political commitment and the structures in place for its implementation and review. In understanding the plausibility and effectiveness of lower emissions development paths, Sathaye et al. (2007) reviewed scenario literature and historical evidence showing that alternative development paths are indeed plausible, and that they are associated with widely different economic, environmental and social consequences. The review also emphasised that lower emissions pathways are not necessarily associated with lower economic growth.

Changing development pathways is not about choosing a mapped out path, but rather about navigating through an uncharted and evolving landscape (Sathaye et al. 2007: 701). This highlights not just the importance of planning, but of review and corrective action in incorporating new information and new goals. The challenge in less developed countries is more of a blank canvas in creating new development visions. It must recognise that resources and institutional capacity are more scarce and that the challenge of poverty eradication is high on the agenda. The challenge in the more industrialised countries is to change existing development paths, one in which lock-in looms large, but resources and institutional capacity are usually higher. However, the flip-side of this coin deserves special attention as changing development paths offers many 'win-wins'. The opportunities in both developed and developing countries include the potential to improve development resilience, enhance new growth, and what could be described as the most important development goal of all, to improve human wellbeing and indeed enhance the environment while seeking transition (O'Mahony 2016).

Thinking About Mitigation and Transition Through the Energy Hierarchy

In articulating the different types of transition the concepts of immaterialisation, dematerialisation and decarbonisation are useful tools described by Tapio et al. (2007: 435–436). *Dematerialisation* refers to the decoupling of a specified environmental harm from material production. Sometimes termed 'eco-efficiency' it includes technical efficiency and technological change.[10] *Decarbonisation* refers to delinking emissions to reduce carbon intensity, it involves moving to 'cleaner' technologies such as renewables and nuclear (low-carbon energy) and fossil fuel substitution by fuel switching to lower carbon fuels such as gas.[11] Dematerialisation and decarbonisation are the largely technical measures, which are necessary, but do not fully address the underlying development path.[12] In that sense they could be considered 'end-of-pipe' (O'Mahony and Dufour 2015a). It is in the first concept of *immaterialisation* that a more fundamental shift is made in a development path. Immaterialisation can refer to a decoupling of material production or consumption from economic growth and human wellbeing.[13] In this approach human wellbeing can be maintained or even advanced while reducing the material consumption that supports it (Jackson 2009; Fleurbaey et al. 2014). More recent research is now attempting to flesh out the concepts and practicalities of how this could be achieved (O'Mahony and Dufour 2015b; O'Mahony 2016) including the MAXWELL project on 'maximising wellbeing and minimising emissions' (Box 3.3).

Box 3.3: 'MAXWELL' on Win-Win Pathways that Improve Wellbeing and Climate Change Mitigation in the EU to 2050

MAXWELL (*maximise wellbeing, minimise emissions*) is an EU Horizon 2020 funded research project[14] at the Finland Futures Research Centre. It addresses a pivotal issue in attempts to reduce greenhouse gas emissions in line with EU legal commitments and intergovernmental treaties. The link between material consumption and climate change is well accepted, increasing global material consumption continues to drive up the greenhouse gas emissions that cause climate change and impacts on sustainability. Approaches to climate change mitigation policy tend to focus on technology and efficiency, but the problem of rising material consumption can often overwhelm these attempts leading to absolute increases in emissions. A debate on the place of reducing material consumption as a means of decreasing emissions and achieving sustainability has been unresolved since the first attempts to implement intergovernmental climate treaties in the 1990s. A perception exists that mitigating climate change, through the deep reductions in emissions required by 2050 involves cost and loss. Such a perception of declines in 'living standards' is an unpopular perception both with the public and with policymakers. However, some researchers have begun to explore the theory of a 'win-win' in mitigation, where overall wellbeing can be maintained or even advanced as emissions are reduced, through a decoupling of human wellbeing from consumption. This has been a controversial topic, and while the theory appears sound, there has been insufficient theoretical or empirical study. There has been little or no policy implemented towards reduced material consumption beyond 'sustainable consumption and production' which returns to technology and efficiency. MAXWELL engages with this prominent gap in the understanding of the wellbeing/climate win-win. Different conceptions of wellbeing are explored, alternative scenarios of wellbeing in the EU to 2050 are created, and modelling seeks to quantify the changes in emissions that arise. Particular attention will be paid to pathways that balance overall wellbeing rather than prioritise material consumption.

One of the possible tools to begin framing the importance of the development path in determining mitigation and future energy transition was articulated by O'Mahony (2016) at a seminar on the low-carbon transition at Princeton Environment Institute's Climate Futures Initiative (see Fig. 3.3). Building on the widely known waste hierarchy, a mainstay of EU waste policy which seeks to place prevention at its heart (EC 2008), the renovated energy hierarchy seeks to clarify that in development and national energy planning, immaterialisation of the development path should always be considered first to reduce demand and emissions at source. It could be

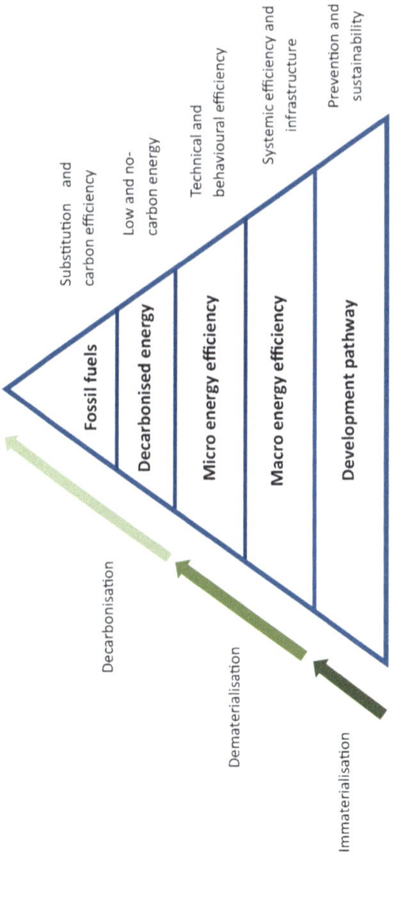

Fig. 3.3 A renovated energy hierarchy for thinking about energy mitigation and transition. Source: O'Mahony (2016)

described as 'designing out energy and emissions'. Technological measures should follow only after a sustainable development pathway has been articulated and implemented. This would enable a lower emissions trajectory, and technology is then used to provide the required energy services and remove and minimise emissions. This is inherently a pathway that is likely to achieve lower cost, lower environmental impacts and more successful mitigation. It can even be dovetailed with win-win synergy outcomes of higher growth and/or improved human wellbeing and environmental quality. Further up the hierarchy, the technical measures dematerialise and decarbonise energy consumption. The hierarchy establishes that transition is a fundamental challenge that addresses the more direct drivers (including the goals of society, economic structure and development policy to spatial, transport and energy-system planning) to the underlying fundamental drivers (social, cultural and institutional) that dictate the orientation of society and politics to sustainability, environmental protection and equity. They could be described as both goal-oriented human development outcomes in terms of quality of life and wellbeing, but also as processes. They dictate the direction that the society and economy will take in moving towards sustainability through its values and ethics, its policymaking style and in the political empowerment of different voices.

From Socio-technical Transitions to Sustainable Development

Some of the literature on innovation and technological change has sought to adopt a broad systemic perspective on how new technologies emerge and diffuse, and illustrates how the technological transition can occur in changing pathways. The changes in technology, their causes and the implications for societies have been actively studied in the social sciences since the late eighteenth century by historians, economists and sociologists (Fleurbaey et al. 2014: 313). Within this literature it is often assumed that technological change is not chaotic but proceeds in certain directions including technological regime (Nelson and Winter 2002) and technological paradigms (Dosi and Nelson 1994). Technological regimes refer to shared beliefs among technicians about what is feasible, whereas technological paradigms refer to the selected set of objects engineers work on, and to the selected set of problems they choose to address. More recent research has yielded two major perspectives on technology transitions in the multi-level perspective on '*socio-technical systems*' of Geels (2002) and the concept of '*technological innovations systems*' of Bergek et al. (2008).

In socio-technical systems a socio-technical landscape corresponds to the regulatory, institutional, physical and behavioural environment in which innovations emerge with considerable inertia at this level. These aspects of technological transition and its social determinants are important contributors to understanding the low-carbon transition from transition management literature. Nevertheless, as they focus on new technologies and systems of supply this is what Shove and Walker (2010: 476) refer to as the 'narrow slice of what is a much wider debate about social systemic change'. Accounting for technology as a fundamental contributor to development it would then be useful to augment such technological transitions with 'societal transitions'. Development paths can then include understanding of the needs, wants and demands of the people. Socio-technical transitions are an aid in understanding the interaction of technology with society, but they do not offer a discussion of where technology is placed within our vision for society or wider social change. It is here that 'systems thinking' and political economy approaches to development in general would be useful to unpack the assumptions. These assumptions are hidden in our economic and technological conceptions of the future, and scenario approaches can prove quite useful in exploring them more.

Amartya Sen offers the capability approach as a critique of income, commodity production, opulence and financial success, as income is the means and not the ends of development (Anand and Sen 2000: 2031). This perspective is widely accepted in development studies in general. Such a human development perspective should also be applied to technology as it can be a useful tool in furthering social progress and improving our lives, but should also be viewed as a means and not the end of development, which is ultimately to live good and useful lives.[15] This critique of the economy and technology, one that largely does not arise within the literature on mitigation and energy, is where sustainable development pathways can begin to fully integrate the concerns of development in general. Such a critique has been suggested for some time but has largely failed to translate into how we analyse the problem of climate mitigation, or propose robust development policies that can lead our societies to lower emissions outcomes and sustainability and wellbeing in general.

Ethics, Values and Policies

In arriving at the place for development in mitigation it is important to note the observation of Nakicenovic et al. (2000: 114) in the IPCC Special Report on Emission Scenarios: 'Beyond the satisfaction of basic needs, the issue of what constitutes "development" involves many cultural, social, and economic dimensions that cannot be resolved by scientific methods, but are inherently a question of values, preferences, and policies.' This places social sciences and political discourse at the heart of defining low-carbon sustainable pathways. It also acknowledges that the choices made cannot be separated from public policy and institutional tasks of balancing the social, environmental and economic dimensions of development, and the equity considerations of how these unfold in a development pathway. The previous discussion has highlighted how technology is necessary but insufficient for transition. It has also been noted that economic instruments such as carbon taxes and other pricing mechanisms are necessary but are unlikely to change behaviour or drive investment at the speed or scale required (Barker et al. 2007b: 662). In addition, economic and technological modelling and analysis can provide useful 'evidence' to support the policymaking process, but they are not substitutes for the politics of decision-making. Modelling is not an entirely 'objective' empirical exercise and there are hidden worldviews and potential biases implicit in each modelling approach (Nielsen and Karlsson 2007: 311).

When the economic system is left to market forces, the approach to development planning can come with large and avoidable social costs (Storm and Naastepad 2007: 1173) as the 2008 financial crisis illustrates. Intervention is controversial and contested and what is judged as government and market failure may be subjective (Shafaeddin 2004). Indeed some political and ideological arguments view intervention as undesirable, as in those who believe that the role of government should be minimised. However, climate change is an example of skewed development and the Stern Report (Stern et al. 2006: Part 1 page 1) noted that climate change is 'the greatest example of market failure we have ever seen'. In the global futures scenarios, falling greenhouse gas emissions are associated with higher government intervention, while rising greenhouse gas emissions are associated with low government intervention (Morita and Robinson 2001: 141). It is not then a question of if one should intervene, but of when, how and how much? The key appears to be an appropriate balance. The words of the late Indian economist Sukhamoy Chakravarty may be instructive in that 'the market is a bad master, but can be a good servant' (Chakravarty 1993: 420). The market and technology can be employed in the pursuit of equitable and sustainable development, but they require guardrails in how they are understood in research and in how they are applied in policy. Rosa and Dietz (2012: 583–584) noted that a country's institutions and culture have been less systematically studied than the prominent greenhouse gas driving forces of population, affluence and technology. They suggest that institutions and culture are important in determining the composition of consumption and the technologies used to support it. A more broad perspective would also emphasise that a country's institutions and culture will determine what kind of intervention and regulation is appropriate, how social equity is determined in development and the balance between social, environmental and economic pillars. Rosa and Dietz also identify 'values, beliefs, norms, trust and world-views' as key drivers of environmental change. With 'an immense literature' examining their role in shaping the environmental behaviour of individuals and the cross-national differences in environmental concerns, they then caution that the 'widely held expectation that such factors influence cross-national differences in environmental stress, and in particular, emissions of greenhouse gases, remains undisciplined by a supporting body of research' (Rosa and Dietz 2012: 584).

The ethics of development have profound consequences for the equity of development between and within countries and also across future generations. Equity issues are central to the debates about development, whether this concerns the sharing of resources locally within a country, or indeed globally. In incorporating a sustainable development perspective, future generations have an ethical claim on the ability to meet their needs and develop their capabilities. This impacts on the development choices that are made today and how resources are consumed or conserved. The environment and a stable climate have a central role to play in this debate. Whether 'anthropocentric' (human-centred) or 'ecocentric' (nature-centred) ethical worldviews are applied to development will affect whether the environment is preserved for the value of the ecosystem services it provides to humans, or for its intrinsic value in and of itself. Some development discourse considers natural capital as substitutable (Anand and Sen 2000), implying that we can trade economic and social capital that is developed now for the loss of natural capital in the future as it is consumed in the process of development. It is now known that there are very different types of natural capital, some of which cannot be replaced or entail significant risks (European Environment Agency 2015: 51). Even if we take only a human-centred focus on the value of the environment, any approach that applies an assumption of fully substitutable natural capital is becoming exceedingly problematic as the issue of climate change ably illustrates. A burgeoning literature places empirical evidence at odds with the theoretical assumption of fully substitutable capital. The UNDP-UNEP Millennium Ecosystem Assessment has identified 15 of 24 ecosystem services that function as humanity's life-support system that are now in serious decline (MEA 2005). Studies such as Steffen et al. (2015) have identified four of nine planetary boundaries[16] that have crossed what is described as the 'safe operating space' that is characteristic of the geological epoch in which human civilisation has emerged. This heightens the need to consider the balance of development with respect to the environment. An additional consideration is that current economic development patterns are also causing problems for social capital in many countries (Bartolini 2014). As the pillars of sustainable development—social, environmental and economic—are interconnected, there are also consequences of failing development patterns for the economy. The ethical considerations are intrinsic in resolving these issues and if there are consistent failures to recognise these, the consequences will be more significant for even the most dogmatic ideological interpretation of the primacy of the economy and growth.

The policy realisation of this is that there is an urgent need to effectively consider policies not in thematic or departmental silos, but as integrated conceptions of development across all government functions and wider governance.[17] The dominant national mitigation policy framework worldwide applies an economic 'sectoral approach' which mostly addresses techno-economic measures in individual sectors, and appears related to the conventions of economic data collection and modelling.[18] This continues to guide policy away from the overall development focus that is known to be required and fails to integrate policy, as indicated by Casado-Asensio and Steurer (2016). The economic, environmental, social, cultural, technological, ethical and governance challenges of development are not separate challenges to be parsed and disaggregated, but must be unified towards a robust vision of what type of world we wish to create. These are interrelated multidimensional problems, but more importantly, they also offer the opportunities of co-benefits and synergies. Integrated policy can address multiple goals while at the same time seeking win-wins. This is the essence of addressing sustainable development pathways and not tagging-on techno-economic solutions at the tail-end. It is more than a 'prevention-not-cure' approach and can enable and empower positive and desirable win-win outcomes. The example of global agriculture and food consumption is prescient. Where diets are shifted away from current patterns of excessive meat consumption towards plant-based options, a whole host of co-benefits and synergies can arise: improved public health outcomes, lower consumer costs, reduced inequality through lower global market competition for resources, new economic opportunities, reductions in pollution, in biodiversity loss and in land-take, improved food security and significant reductions in greenhouse gas emissions and other environmental pollutants and pressures. The public health, economic and environmental costs of continuing to publicly subsidise livestock agriculture are therefore grossly at odds with sustainable development and enhancing human and environmental wellbeing. The key to unlocking this pattern may be in how to support the transition and adaptation of economic activity and livelihoods away from meat and dairy in countries where these industries are prioritised.

There are potential trade-offs in determining where transition efforts in general fit with other societal goals, but it is in integrated policy and development pathways that these can be managed. The Technical Summary of the IPCC Fifth Assessment Report Working Group III provides a succinct summary of the place of synergies and trade-offs through sustainable development:

> Mitigation efforts generate tradeoffs and synergies with other societal goals that can be evaluated in a sustainable development framework. The many diverse goals that societies value are often called 'sustainable development'. A comprehensive assessment of climate policy therefore involves going beyond a narrow focus on distinct mitigation and adaptation options and their specific co-benefits and adverse side-effects. Instead it entails incorporating climate issues into the design of comprehensive strategies for equitable and sustainable development at regional, national, and local levels. Maintaining and advancing human well-being, in particular overcoming poverty and reducing inequalities in living standards, while avoiding unsustainable patterns of consumption and production, are fundamental aspects of equitable and sustainable development. Because these aspects are deeply rooted in how societies formulate and implement economic and social policies generally, they are critical to the adoption of effective climate policy. (Edenhofer et al. 2014: 40)

Conclusion

In Chap. 3 we have moved from framing the problem to characterising how this can be understood in the low-carbon transition. The mitigation of greenhouse gas emissions continues to focus on economic and technological measures, but these are inadequate in the context of the need to transition to sustainable development pathways. If we are to meet deep emissions reductions targets, and achieve economically, socially and environmentally sustainable outcomes, a development focus is necessary that addresses underlying social, cultural and governance drivers. The limitations of techno-economic approaches to mitigation have been evident for some time, and yet there has been a practice inertia in moving towards more holistic forms of understanding transition and policymaking for its realisation. Casado-Asensio and Steurer (2016: 101) have noted that national mitigation strategies have rarely been studied, but are not delivering much by way of emissions reductions describing them as 'lacklustre bookkeeping'. They are failing to integrate, to manage conflicts or develop synergies in a 'lose-lose' rather than a 'win-win' approach. Sathaye et al. (2007) have reviewed historical evidence and scenario literature showing that alternative development paths are indeed plausible, and can deliver more desirable outcomes. In implementing national development pathways and integrated climate policy, a planning, monitoring, checking and corrective action cycle is necessary for effective policy.

The human drivers influence not only the technological transition but the very nature of the development path in each country. Implementing a sustainable development path has the aim of addressing underlying drivers and delivering a lower emissions trajectory before 'climate policy' is implemented. Implementing climate policy without a sustainable development pathway is essentially an end-of-pipe response (O'Mahony and Dufour 2015b: 418). Beginning with a sustainable development pathway opens up opportunities for synergies and win-wins between different policy goals. The social dimensions of a sustainable development path are strong, both as outcomes and processes, and require much greater attention. The place of ethics, values and politics cannot be fully separated. In incorporating these perspectives, the opportunity to radically reduce emissions and improve human wellbeing and the environment can materialise in more balanced development.

NOTES

1. The human sources of greenhouse gas emissions that are now causing climate change include the burning of fossil fuels: coal, oil, gas and peat for energy; in agriculture animal rearing is a significant source; and in manufacturing there are emissions from industrial processes including cement production. Other notable sources include waste, land use change and deforestation.
2. Low-carbon energy technology could be in the form of renewables such as solar, wind, hydro and biomass, nuclear energy, or the newer idea of carbon capture and storage where carbon emissions from the burning of fossil fuels are captured and buried. An even more recent idea seeks to burn biomass, and then capture and bury the carbon that arises to physically reduce the quantities of carbon emissions in the atmosphere.
3. $I = P \times A \times T$, where I is the environmental impact, P is population, A is affluence and T is technology.
4. Where 'ppm' is parts per million of CO_2 in the atmosphere, as a measure of the concentration of the main greenhouse gas. CO_2 levels in the atmosphere have risen from 280 ppm before the industrial revolution to 405 ppm in December 2016 according to the National Oceanic & Atmospheric Administration in the US.
5. Of what technology is deployed, what type of economy and economic instruments are implemented and how technology and the economy relate.
6. Usually defined as policy that reduces emissions from the consumption of fossil fuels through technical means such as efficiency, renewables and, as an economic measure, carbon-pricing by means of taxes. It can also include reducing other greenhouse gases from deforestation, agriculture and industrial processes.
7. Where there are additional benefits to emissions mitigation such as reductions in air pollution and improvements in human health through burning less fossil fuels.
8. Where choosing a pathway leads to 'win-wins' as synergies across different policy domains. An example is reducing the over-consumption of meat in affluent countries which would contribute to public health objectives by reducing disease, but also to reduced GHG emissions at the same time.
9. Nature conservation, legal frameworks, property rights, rule of law, taxes and regulation, production, security and safety of food, consumption patterns, human and institutional capacity building efforts, R&D, financial schemes, technology transfer, energy efficiency and energy options, see Barker et al. (2007a: 33).
10. To reduce the intensity of emissions per unit production or per unit economic output.

11. This reduces the intensity of emissions per unit of energy.
12. They may have some limited effect but they do not directly address the development path.
13. A simple example is to use dense spatial planning; this approach tends to reduce transport energy consumption due to less travel requirements. Co-benefits include less spending on transport energy and improved public health and air quality with no costs to the economy or to wellbeing but with benefits.
14. This project is funded by the European Commission research and innovation programme under the Marie Skłodowska-Curie grant agreement No. 657865. For further information see www.utu.fi/en/units/ffrc/Pages/home.aspx or http://sdfutures.fi/
15. And also how we live in relation to the Earth's ecosystems for 'environmental wellbeing'. Environmental wellbeing has its own intrinsic value in addition to its contribution to human wellbeing.
16. These are: extinction rate, deforestation, atmospheric CO_2 and the flow of nitrogen and phosphorus, see Steffen et al. (2015).
17. State, market and civil society.
18. United Nations Framework Convention on Climate Change guidance on 'national communications' and 'progress reports' has also guided policy towards a sectoral approach (UNFCCC 2008).

References

Anand, S., and A. Sen. 2000. Human Development and Economic Sustainability. *World Development* 28 (12): 2029–2049.

Banuri, T., J. Weyant, G. Akumu, A. Najam, L. Pinguelli Rosa, S. Rayner, W. Sachs, R. Sharma, and G. Yohe. 2001. Setting the Stage: Climate Change and Sustainable Development. In *Climate Change 2001: Mitigation, Report of Working Group III, Intergovernmental Panel on Climate Change (IPCC)*, ed. B. Metz, O. Davidson, R. Swart, and J. Pan. Cambridge: Cambridge University Press.

Barker, T., I. Bashmakov, L. Bernstein, J.E. Bogner, P.R. Bosch, R. Dave, O.R. Davidson, et al. 2007a. Technical Summary. In *Climate Change 2007: Mitigation. Contribution of Working Group III to the Fourth Assessment Report of the Intergovernmental Panel on Climate Change*, ed. B. Metz, O.R. Davidson, P.R. Bosch, R. Dave, and L.A. Meyer. Cambridge: Cambridge University Press.

Barker, T., I. Bashmakov, A. Alharthi, M. Amann, L. Cifuentes, J. Drexhage, M. Duan, et al. 2007b. Mitigation from A Cross-sectoral Perspective. In *Climate Change 2007: Mitigation. Contribution of Working Group III to the Fourth Assessment Report of the Intergovernmental Panel on Climate Change*, ed. B. Metz, O.R. Davidson, P.R. Bosch, R. Dave, and L.A. Meyer. Cambridge: Cambridge University Press.

Bartolini, S. 2014. Building Sustainability Through Greater Happiness. *The Economic and Labour Relations Review* 25 (4): 587–602.

Bergek, A., S. Jacobsson, B. Carlsson, S. Lindmark, and A. Rickne. 2008. Analyzing the Functional Dynamics of Technological Innovation Systems: A Scheme of Analysis. *Research Policy* 37: 407–429.

Boström, M. 2012. A Missing Pillar? Challenges in Theorizing and Practicing Social Sustainability: Introduction to the Special Issue. *Sustainability: Science, Practice and Policy* 8 (12): 3–14.

Calverley, D., R. Wood, S. Mander, K. Anderson, S. Glynn and F. Nicholls. 2009. *Towards a 2°C Future: Emission Reduction Scenarios for Wales*. A research report by The Tyndall Centre at the University of Manchester, commissioned by the Climate Change Commission of the Welsh Assembly Government. Accessed 24 March 2017. http://gov.wales/docs/desh/publications/111014towardsa2en.pdf

Casado-Asensio, J., and R. Steurer. 2016. Bookkeeping Rather than Climate Policy Making: National Mitigation Strategies in Western Europe. *Climate Policy* 16 (1): 88–108.

Chakravarty, S. 1993. *Selected Economic Writings*. Delhi: Oxford University Press.

Dosi, G., and R.R. Nelson. 1994. An Introduction to Evolutionary Theories in Economics. *Journal of Evolutionary Economics* 4: 153–172.

EC. 2008. Directive 2008/98/EC of the European Parliament and of the Council of 19 November 2008 on Waste and Repealing Certain Directives. *Official Journal of the European Communities* L312, 51: 3–30. http://eur-lex.europa.eu/JOHtml.do?uri=OJ:L:2008:312:SOM:EN:HTML

Edenhofer, O., R. Pichs-Madruga, Y. Sokona, S. Kadner, J.C. Minx, S. Brunner, S. Agrawala, et al. 2014. Technical Summary. In *Climate Change 2014: Mitigation of Climate Change. Contribution of Working Group III to the Fifth Assessment Report of the Intergovernmental Panel on Climate Change*, ed. O. Edenhofer, R. Pichs-Madruga, Y. Sokona, E. Farahani, S. Kadner, K. Seyboth, A. Adler, I. Baum, S. Brunner, P. Eickemeier, B. Kriemann, J. Savolainen, S. Schlömer, C. von Stechow, T. Zwickel, and J.C. Minx. Cambridge: Cambridge University Press.

Ehrlich, P.R., and J.P. Holdren. 1971. Impact of Population Growth. *Science* 171: 1212–1217.

European Environment Agency. 2015. *The European Environment: State and Outlook 2015*. Copenhagen: European Environment Agency. http://www.eea.europa.eu/soer

Fisher, B.S., N. Nakicenovic, K. Alfsen, J. Corfee Morlot, F. de la Chesnaye, J.-Ch. Hourcade, K. Jiang, et al. 2007. Issues Related to Mitigation in the Long Term Context. In *Climate Change 2007: Mitigation. Contribution of Working Group III to the Fourth Assessment Report of the Intergovernmental Panel on Climate Change*, ed. B. Metz, O.R. Davidson, P.R. Bosch, R. Dave, and L.A. Meyer. Cambridge: Cambridge University Press.

Fleurbaey, M., S. Kartha, S. Bolwig, Y.L. Chee, Y. Chen, E. Corbera, F. Lecocq, et al. 2014. Sustainable Development and Equity. In *Climate Change 2014: Mitigation of Climate Change. Contribution of Working Group III to the Fifth Assessment Report of the Intergovernmental Panel on Climate Change*, ed. O. Edenhofer, R. Pichs-Madruga, Y. Sokona, E. Farahani, S. Kadner, K. Seyboth, A. Adler, I. Baum, S. Brunner, P. Eickemeier, B. Kriemann, J. Savolainen, S. Schlömer, C. von Stechow, T. Zwickel, and J.C. Minx. Cambridge: Cambridge University Press.

Geels, F. 2002. Technological Transitions as Evolutionary Reconfiguration Processes: A Multi-level Perspective and a Case-study. *Research Policy* 31: 1257–1274.

Halsnæs, K., P. Shukla, D. Ahuja, G. Akumu, R. Beale, J. Edmonds, C. Gollier, et al. 2007. Framing Issues. In *Climate Change 2007: Mitigation. Contribution of Working Group III to the Fourth Assessment Report of the Intergovernmental Panel on Climate Change*, ed. B. Metz, O.R. Davidson, P.R. Bosch, R. Dave, and L.A. Meyer. Cambridge, UK and New York, NY: Cambridge University Press.

IPCC. 2012. *Managing the Risks of Extreme Events and Disasters to Advance Climate Change Adaptation*. Intergovernmental Panel on Climate Change. Cambridge: Cambridge University Press.

———. 2014. Summary for Policymakers. In *Climate Change 2014: Mitigation of Climate Change. Contribution of Working Group III to the Fifth Assessment Report of the Intergovernmental Panel on Climate Change*, ed. O. Edenhofer, R. Pichs-Madruga, Y. Sokona, E. Farahani, S. Kadner, K. Seyboth, A. Adler, I. Baum, S. Brunner, P. Eickemeier, B. Kriemann, J. Savolainen, S. Schlömer, C. von Stechow, T. Zwickel, and J.C. Minx. Cambridge: Cambridge University Press.

Jackson, T. 2009. *Prosperity Without Growth? The Transition to a Sustainable Economy*. London: The Sustainable Development Commission. Accessed 26 July 2017. http://www.sd-commission.org.uk/data/files/publications/prosperity_without_growth_report.pdf

Kaya, Y. 1990. Impact of Carbon Dioxide Emission Control on GNP Growth: Interpretation of Proposed Scenarios. Paper presented to the *IPCC Energy and Industry Subgroup, Response Strategies Working Group*, Paris (Mimeo).

McGillivray, M. 2007. *Human Well-Being: Concept and Measurement*. Basingstoke: Palgrave Macmillan.

MEA (Millennium Ecosystem Assessment). 2005. *Ecosystems and Human Well-being: Current State and Trends*. Vol. 1. Washington, DC: Island Press.

Metz, B., M. Berk, M. den Elzen, B. de Vries, and D. van Vuuren. 2002. Towards an Equitable Climate Change Regime: Compatibility with Article 2 of the Climate Change Convention and the Link with Sustainable Development. *Climate Policy* 2: 211–230.

Metz, B., O. Davidson, R. Swart, and J. Pan. 2001. Climate Change 2001: Mitigation. In *Third Assessment Report of the Intergovernmental Panel on Climate Change*, ed. B. Metz et al. New York: Cambridge University Press.

Morita, T., and J. Robinson. 2001. Greenhouse Gas Emission Mitigation Scenarios and Implications. In *Climate Change 2001: Mitigation*, ed. B. Metz, O. Davidson, R. Swart, and J. Pan, 117–166. Cambridge: Cambridge University Press.

Nakicenovic, N., J. Alcamo, G. Davis, B. de Vries, J. Fenham, S. Gaffin, K. Gregory, et al. 2000. *Special Report on Emissions Scenarios. Working Group III, Intergovernmental Panel on Climate Change (IPCC)*. Cambridge: Cambridge University Press.

Nelson, R.R., and S.G. Winter. 2002. Evolutionary Theorizing in Economics. *The Journal of Economic Perspectives* 16: 23–46.

Nielsen, S.K., and K. Karlsson. 2007. Energy Scenarios: A Review of Methods, Uses and Suggestions for Improvement. *International Journal Global Energy Issues* 27 (3): 302–322.

O'Mahony, T. 2016. The Concept of "Wellbeing" as Driver of Future Mitigation. Sustainable well-being seminar at the Climate Futures Initiative, Princeton University, Princeton, NJ, USA, 11 February 2016.

O'Mahony, T., and J. Dufour. 2015a. Tracking Development Paths: Monitoring Driving Forces and the Impact of Carbon-free Energy Sources in Spain. *Environmental Science and Policy* 50: 62–73.

O'Mahony, T., and J. Dufour. 2015b. The Social and Cultural Dimensions of Sustainable Development, Mitigation and Scenarios: Grasping the Opportunities for Human Development. In *Sustainable Futures in a Changing Climate*, ed. Aino Hatakka and Jarmo Vehmas. Proceedings of the Conference *Sustainable Futures in a Changing Climate*, 11–12 June 2014, Helsinki, Finland. FFRC eBOOK 2/2015. Finland Futures Research Centre, University of Turku, 414–427.

Reed, D., ed. 1996. *Structural Adjustment, The Environment, and Sustainable Development*. London: Earthscan.

Rogner, H.-H., D. Zhou, R. Bradley, P. Crabbé, O. Edenhofer, B. Hare, L. Kuijpers, and M. Yamaguchi. 2007. Introduction. In *Climate Change 2007: Mitigation. Contribution of Working Group III to the Fourth Assessment Report of the Intergovernmental Panel on Climate Change*, ed. B. Metz, O.R. Davidson, P.R. Bosch, R. Dave, and L.A. Meyer. Cambridge: Cambridge University Press.

Rosa, E.A., and T. Dietz. 2012. Human Drivers of National Greenhouse Gas Emissions. *Nature Climate Change* 2: 581–556.

Sathaye, J., A. Najam, C. Cocklin, T. Heller, F. Lecocq, J. Llanes-Regueiro, J. Pan, et al. 2007. Sustainable Development and Mitigation. In *Climate Change 2007: Mitigation. Contribution of Working Group III to the Fourth*

Assessment Report of the Intergovernmental Panel on Climate Change, ed. B. Metz, O.R. Davidson, P.R. Bosch, R. Dave, and L.A. Meyer. Cambridge: Cambridge University Press.

Schopenhauer, A. 1966. *Le monde comme volonté et comme representation*. Paris: Presses Universitaires de France, original 1819.

Sen, A. 1985. *Commodities and Capabilities*. Amsterdam: Elsevier.

———. 1992. *Inequality Reexamined*. New York: Russell Sage Foundation.

Shafaeddin, S.M. 2004. *Who is the Master? Who is the Servant? Market or Government? An Alternative Approach: Towards a Coordination System*. United Nations Conference on Trade and Development. Discussion Papers No. 175, August 2004. Accessed 28 March 2010. http://www.unctad.org/en/docs/osgdp20049_en.pdf

Shove, E., and G. Walker. 2010. Governing Transitions in the Sustainability of Everyday Life. *Research Policy* 39: 471–476.

Steffen, W., K. Richardson, J. Rockstrom, S.E. Cornell, I. Fetzer, E.M. Bennett, R. Biggs, et al. 2015. Planetary Boundaries: Guiding Human Development on a Changing Planet. *Science* 347 (6223): 1259855.

Stern, D.I. 2004. The Rise and Fall of the Environmental Kuznets Curve. *World Development* 32 (8): 1419–1439.

Stern, N., S. Peters, V. Bakhshi, A. Bowen, C. Cameron, S. Catovsky, D. Crane, et al. 2006. *Stern Review: The Economics of Climate Change*. London: HM Treasury.

Stiglitz, J.E., A. Sen, and J. Fitoussi. 2009. *Report by the Commission on the Measurement of Economic Performance and Social Progress*. Paris: Commission on the Measurement of Economic Performance and Social Progress.

Storm, S., and C.W.M. Naastepad. 2007. Sukhamoy Chakravarty: The Feasibility of Equitable Growth. *Development & Change* 38 (6): 1173–1185.

Swart, R., J. Robinson, and S. Cohen. 2003. Climate Change and Sustainable Development: Expanding the Options. *Climate Policy*, Special Issue on *Climate Change and Sustainable Development* 3 (S1): S19–S40.

Tapio, P., D. Banister, J. Luukkanen, J. Vehmas, and R. Willamo. 2007. Energy and Transport in Comparison: Immaterialisation, Dematerialisation and Decarbonisation in the EU15 Between 1970 and 2000. *Energy Policy* 35: 433–451.

UNFCCC. 2008. *UNFCCC Resource Guide for Preparing the National Communications of Non-Annex I Parties*. Bonn: United Nations Framework Convention on Climate Change.

Unruh, G.C. 2002. Escaping Carbon Lock-in. *Energy Policy* 30 (4): 317–325.

PART II

Development Pathways and the Low-Carbon Future

CHAPTER 4

Development Models: Lessons from International Development

Introduction

In addressing pathways to a low-carbon society, the IPCC's Fifth Assessment Report places adaptation and mitigation in the context of what it labels 'equitable and sustainable development'. In so doing, it acknowledges two important realities: firstly, that 'limiting the effects of climate change is necessary to achieve sustainable development and equity, including poverty eradication' and, secondly, that 'effective decision-making to limit climate change and its effects can be informed by a wide range of analytical approaches', recognising the importance of governance, ethical dimensions, value judgments, economic assessments and 'diverse perceptions and responses to risk and uncertainty' (IPCC 2014: 76). In stating this, the world's scientists (including social scientists whose work is also reviewed as part of the IPCC process) are therefore linking the wicked problem of climate change to the project of international development that emerged in the immediate post-World War II period when the goal of raising living standards in the so-called developing world was established (Rist 1997).

Yet, while alluding to more than half a century of development efforts around the world, the IPCC does not interrogate what lessons this may have to offer the task of transitioning to a low-carbon society. Indeed, the detail with which technological and scientific issues are treated when

P. Kirby, T. O'Mahony, *The Political Economy of the Low-Carbon Transition*, International Political Economy Series,
DOI 10.1007/978-3-319-62554-6_4

considering mitigation and adaptation contrasts with the vague generalities used in referring to development pathways.[1] For example, in stating that 'development along current global pathways can contribute to climate risk and vulnerability, further eroding the basis for sustainable development' (IPCC 2014: 90), the IPCC is raising issues of crucial importance but it does not seek to critique them. This statement implies that current models of development are part of the problem of climate change and its impact, while a more sustainable form of development is possible. Furthermore, the IPCC makes clear that restricting responses to incremental changes to existing systems is inadequate and that 'transformational change' should be considered. As examples of transformational change it mentions 'introduction of new technologies or practices, formation of new financial structures or systems of governance, adaptation at greater scales or magnitudes and shifts in the location of activities'. Building adaptive capacity 'can involve complex governance challenges and new institutions and institutional arrangements', it states (ibid.: 80).

A failure to examine thoroughly what development pathways may be contributing to climate change, what new social, political and economic institutions and institutional arrangements may be required to transition to a low-carbon society, what types of new financial structures can underpin a successful transition, and what shifts of what magnitudes in the location of what activities may be necessary, is a major weakness of our collective understanding of what is involved in low-carbon transition. It both results from the dominance of a particular socio-technological paradigm as identified in Chap. 2, and neglects rich academic and practice-based literatures that research, reflect on and debate pathways to a sustainable and equitable society (for an overview, see Nederveen Pieterse 2001). The consequences of this can be highlighted by referring to a comment by a senior Indian scientist, Aromar Revi, director of the Indian Institute for Human Settlements and an author of the IPCC Fifth Assessment Report. Reflecting on that report, he wrote:

> Greater clarity and innovation is required to identify an effective suite of initiatives that could connect-the-dots with each region's often unique set of choices to enable multiple transitions along future climate-resilient development pathways. This is clearly possible. It requires us to make a series of difficult collective choices. (Revi 2014)

This clarity and innovation is only possible if we draw on the lessons of development theory and practice. This is the purpose of this chapter.

In the next section, it introduces the concept of development pathways, drawing on development theory and practice to help clarify the developmental choices now facing humanity as we attempt to transition to a low-carbon society. Particular attention is paid to debates on the concept of 'sustainable development' that emerged to international recognition in the early 1970s. The next section introduces the concept of development models, as a major contribution to understanding particular development pathways, how they are constituted and their social outcomes. This focuses attention on the fact that development pathways, in their socio-economic sense, coalesce around models with particular state-market-civil society 'governance' relationships that constitute them. This conceptual framework is applied to the challenges of climate change in the following section, elaborating on the different political economy models that have emerged to address the challenges of climate change. The conclusion draws out what can be learnt to inform pathways towards a low-carbon society from critical consideration of political economy models.

Lessons of Half a Century of Development Efforts

The project of international development emerged out of the optimism of the allied victory in the Second World War and the success of the Marshall Plan in helping rebuild war-torn Europe. In his inauguration address in January 1949, US President Harry Truman announced the project of 'making the benefits of our scientific advances and industrial progress available for the improvement and growth of underdeveloped areas', distinguishing this 'programme of development' from 'the old imperialism—exploitation for foreign profit' (quoted in Rist 1997: 250–251). While this project bore traces of earlier attempts at planned economic and social progress, such as colonial economics as practised by British colonial administrators and in Soviet central planning, what was new was the application of the vocabulary of development, essentially a biological concept, to the tasks of planned socio-economic change. Suddenly whole swathes of the world—Africa, Asia, Latin America, the Caribbean, the Pacific—were labelled developing or underdeveloped and placed on a common pathway. As Sachs put it: 'For the first time, the new worldview was thus announced: all the peoples of the earth were to move along the same track and aspire to only one goal—development' (quoted in Kirby 1997: 30). This quickly grew into a significant sector of activity with transnational agencies such as the World Bank and the UN Development Programme

(UNDP), national development programmes in many Western states, NGOs such as Oxfam, and development studies departments such as the UK's Overseas Development Institute (ODI) being established. Working at the ODI in London in the early 1960s, Teresa Hayter shared the optimistic belief that more foreign aid would help improve the situation of the poor in developing countries. 'There was little attempt, however, to define development. Instead, there was an unquestioned assumption among the staff that "development", whatever it was, would lead to improvement in the situation of poor people' (Hayter 2005: 89).

The singular development pathway thus established soon fragmented into many. While the first decade of development in the 1950s placed the emphasis on economic development through Western aid and technology, by the early 1960s social concerns came to the fore as researchers uncovered that a decade of development aid, far from reducing poverty, had worsened the gap between rich and poor. In 1973, researchers Adelman and Morris admitted to being shocked by their research results which found that 'the process of economic modernization shifts the income distribution in favour of the middle class and upper income groups and against lower income groups [so that] the dynamics of economic development appear to work against the poor' (Adelman and Morris 1973: 188). This resulted in attention being focused on how growth could be made to reduce poverty and income inequality, such as policies to extend basic education and to augment labour demand, what was labelled 'Redistribution with Growth'.

As Western researchers were discovering the limitations of an exclusive focus on economic growth, new pathways of development were being forged as colonial powers granted independence to colonies in Asia, Africa and the Caribbean from the 1950s onwards. At the celebrated Bandung conference in Indonesia in 1955, attended by newly independent countries such as India, Pakistan, Burma, Sri Lanka (then named Ceylon), Egypt and Lebanon, the term 'third world' was coined to distinguish this group of countries from the capitalist West and the Soviet East, emphasising their attempts to foster their own pathways to development. The Non-Aligned Movement emerged from a follow-up conference in Belgrade in 1961. This so-called third way[2] to development found expression in a range of different developmental regimes that emerged in the 1960s and 1970s, drawing on national traditions and value systems. These included various forms of African socialism, notably the *ujamaa* (familyhood in Swahili) village-based socialism of Julius Nyerere (President of Tanzania,

1964–1985) and the *négritude* philosophy of Léopold Senghor (President of Senegal, 1960–1980); and the Arab socialism of Gamal Abdel Nasser (President of Egypt, 1956–1970) finding expression in the short-lived United Arab Republic of Syria and Egypt from 1958 to 1961. In the communist bloc, Josip Broz Tito who ruled Yugoslavia from 1943 to 1980 championed independent roads to socialism, broke with Stalin and implemented a self-management system. He was also a leading figure in the Non-Aligned Movement. In the Caribbean, Michael Manley's period as Prime Minister of Jamaica (1972–1980) was marked by a series of socialist reforms while the brief socialist government of Maurice Bishop in Grenada (1979–1983) ended with his execution and a subsequent invasion by US troops. Meanwhile, a number of military socialist regimes took power in various Latin American countries. The government of General Juan Velasco Alvarado in Peru (1968–1975) introduced extensive nationalisation of industries and services, handed leading media over to the control of trade unions, mobilised shanty town dwellers and radically reformed education. The military regimes of Generals Omar Torrijos in Panama (1972–1981), Guillermo Rodriguez Lara in Ecuador (1972–1976) and Juan J. Torres in Bolivia (1970–1971) similarly broke the mould. However, the leading example of a democratic road to socialism was the popularly elected socialist government of Salvador Allende in Chile, elected in 1970 and overthrown by the military in 1973.

Third-way pathways to development remained the exception however, and few lasted very long. Instead the world of development came to be characterised by two competing conceptions, the mainstream capitalist approach labelled modernisation and the critical approach, influenced by but distinct from Marxism, labelled dependency. Modernisation emerged from US universities in the 1950s and promoted a view of development as a move from traditional to modern societies. Associated with theorists such as the sociologist Talcott Parsons (1902–1979) who elaborated the stages through which societies pass on their way to modernity, and the economist Walt Rostow (1916–2003) who outlined five stages through which countries pass to achieve economic 'take-off', modernisation theory accepted Western, democratic, free market, mass consumer societies as the objective towards which all other countries should develop. This was seen as an evolutionary process to be achieved by aid and know-how from developed countries in the expectation that the benefits would 'trickle down' to the poorest of society, something which empirical evidence rarely showed to be the case. However, from the 1950s this came to be challenged by voices from within the so-called developing world. The founding of the UN Economic Commission for Latin America (CEPAL after its initials in Spanish) in 1947 (against opposition from the US) established what Kay called 'the first genuine Third World development school' and it became the intellectual centre for elaborating alternative theories of development to counter those from Western universities (Kay 1989: 26). CEPAL's theory of unequal development challenged the dominant theory of comparative advantage that locked countries into primary commodity exporting; instead CEPAL advocated a state-led industrialisation drive that came to be known as Import-Substitution Industrialisation (ISI), a policy prescription that had major influence among policy makers throughout the developing world (Munck 2013: 114–118). This spurred the emergence of dependency theory as a direct challenge to modernisation, arguing that it was the economic and financial dependence of developing countries on the West that was the main cause of their underdevelopment; profits were repatriated back to developed countries while balanced forms of national development to serve local needs were not possible within this dependent relationship. Andre Gunder Frank (1929–2005) posited a chain-like metropolis-satellite relationship that kept developed countries underdeveloped and he advocated de-linking from the capitalist system as a precondition for development. More nuanced was the work of Fernando Henrique Cardoso (1931–) who served as President of Brazil (1995–2003) and Enzo Faletto (1935–2003) who wrote of 'situations of dependency' that trapped peripheral countries limiting their freedom of action (Cardoso and Faletto 1979: xxiii). Recognising the great differences in levels of development within Latin America, they identified the different strategies of local class actors in achieving different development paths for their countries. Over this period, developing countries promoted through the UN system a New International Economic Order (NIEO), advocating increased and better quality aid, tying multinational investors to codes of conduct, and compensation schemes for developing country producers if commodity prices fell below a certain level.

Though positing very different pathways to development, it was clear by the early 1980s that neither modernisation nor dependency was proving a reliable guide to countries' development. The second half of the 1970s was marked by crisis, due both to the limitations of some of the state-directed and inward-oriented development paths taken by many developing countries, but also due to the impact of the quadrupling of oil prices in 1973 and further rises in 1979. Not only did this kill off any consideration of an NIEO, but countries that had borrowed heavily to support their development efforts now found themselves trapped in a severe problem of indebtedness (Di Muzio and Robbins 2016: 75–85). This greatly limited the extent to which they could follow their own paths as they became dependent on agreeing structural adjustment packages with the IMF and the World Bank, liberalising trade and investment rules, privatising nationalised industries and utilities, and downsizing the state. As Kiely puts it: 'This was the start of the debt crisis and with it, the shift to neo-liberal policies in the developing world' (Kiely 2007: 68). Associated with the leadership of Margaret Thatcher in the UK and Ronald Reagan in the US, and codified as the Washington Consensus, this opened a period characterised by the retreat of the state and the liberalisation of market regulation, deepening greatly the inroads of multinational corporations throughout the developing world (Payne 2005: 73–79). Faced with this determined return to mainstream neo-classical economic theories, development retreated downwards, focusing on 'basic needs', an approach to development that sought to ensure standards of education, health and housing at grassroots level. For Stewart this was a 'minimalist approach' (Stewart 1985: 4), and it was seen by some experts that, in this new situation, 'the idea of development … is falling apart and in danger of total collapse'. It is 'scattered by the winds of change over a wide terrain of intellectual enquiry, making the task of synthesis *a priori* impossible' (Hoogvelt 1997: x, xi). Various approaches towards rescuing development as a distinct terrain of intellectual inquiry and practical endeavour emerged in response, most notably the concept of sustainable development, referred to widely in the discussions of pathways to a low-carbon society (see Box 4.1), and human development, an attempt at a new synthesis that emerged from the UNDP from 1990 onwards (Alkire 2010).

Box 4.1: Sustainable Development: 'A Bundle of Neat Fixes'?

The pressure of development on the environment was being recognised by colonial administrators in the nineteenth century but it was only in the 1970s that it became the focus of intellectual and political action. The first UN conference on the human environment took place in Stockholm in 1972, 20 years before the famous Earth summit in Rio de Janeiro in 1992 that brought the issue to public and political prominence. However, it was the report of the World Commission on Environment and Development (WCED) in 1987, entitled *Our Common Future* but referred to as the Brundtland report after the chair of the Commission, former Norwegian prime minister Gro Harlem Brundtland, that popularised the term 'sustainable development'. This gave us the much-quoted definition of sustainable development as 'development that meets the needs of the present without compromising the ability of future generations to meet their own needs' (WCED 1987: 43). However, as Baker points out, the concept also contains within it two key elements: the essential needs of the world's poor on the one hand and, on the other, the limits posed by the environment to meeting these needs (Baker 2006: 20). This is seen by Robinson as 'the radical aspect' of Brundtland, namely 'that ecological sustainability cannot be achieved if the problem of poverty is not successfully addressed around the world' (Robinson 2004: 372).

Instantly taken up by government and industry, the concept of 'sustainable development' was controversial from the beginning. Lélé wrote of it as 'a bundle of neat fixes ... that will unite everybody from the profit-minded industrialist and risk-maximising subsistence farmer to the equity-seeking social worker, the pollution-concerned or wildlife-loving First Worlder, the growth-maximising policy maker, the goal-oriented bureaucrat, and therefore, the vote-counting politician' (Lélé 1991: 613). Concerns were raised about how the concept might be used by government and business to promote a cosmetic environmentalism, allowing growth to proceed unchecked. As a result, many NGOs and environmentalists prefer the term 'sustainability' (Robertson 2014). However, Robinson has argued that perhaps its greatest impact has been to spur the development of sustainability standards and certification for products and services. Baker sees it as representing 'a radical agenda for social change' though one that has been restricted by 'international political and economic processes' (Baker 2006: 218).

Beyond debates on the concept itself, lie deeper issues that are often obscured. At heart these relate to different philosophical and moral conceptions of the relationship of humanity to one another, to future generations and to nature, and to how this should be organised. As Robinson writes, it means that achieving a sustainable society 'is not fundamentally a scientific or technical issue' but one that raises fundamental questions about our values and about how we organise society. It is necessary then 'to address profound issues of opportunity, distribution, material needs, consumption and empowerment', which in turn raise 'important issues of social and political organization and governance'. He adds that 'these issues are likely to be much more intractable than those related to achieving improvements in eco-efficiency' (Robinson 2004: 379). However, if this is the promise of the concept of 'sustainable development' there is little evidence that it has spurred any significant consideration of these issues among decision-makers and the wider public in elaborating pathways towards a low-carbon society.

By the 1990s it was widely considered that the development project was in crisis (Selwyn 2014), overtaken by the globalisation of trade, financial flows, production chains and human (though not labour) mobility. The world of developing countries fragmented into the 'emerging economies', some surpassing the dynamism of Western economies, many achieving economic growth but within very unequal societies, and not a small number mired in conflict, breakdown and even collapse. While in crisis, the development project provides us with many lessons about how we can plan social change from local to global levels, lessons that are essential to inform us as we move urgently into a new type of 'development' project, namely how to transition to a low-carbon society. Before drawing these lessons, however, the narrative of half a century of development efforts tells us nothing about the structuring of power relations that underpinned developmental success when achieved, or that help account for its lack of success. This structuring of power relations between state, market and society we refer to as political economy models of development.

Political Economy Models of Development

In his survey of trends in development theory from the 1950s to the early 2000s, Nederveen Pieterse makes the point that 'development thinking has been, more or less successively, state-led (classical political economy, modernization, dependency), market-led (neoliberalism) and society-led (alternative development)'. This recognises that all development thinking rests on political economy configurations, namely on particular interrelationships between state, market and society. He adds that 'it is increasingly understood that development action needs *all* of these in new combinations' (Nederveen Pieterse 2001: 17; emphasis in original). This applies equally to actions to enable countries to transition to a post-carbon society. The crucial question, therefore, is *what* combinations of state, market and society might best enable that transition to take place. Examining the different combinations of state, market and society that emerged in the attempts to transition countries to development can offer lessons.

As discussed in Chap. 2, political economy models rest on ideal types (see Table 2.2). This heuristic device may help to distinguish the central actors in particular models but it fails to appreciate that models emerge from social struggles and reflect institutional, cultural, economic and social factors distinctive to different societies. Yet, for all their distinctiveness, models tend to emerge in regional clusters, indicating that certain common features distinguish them in different regions of the world. A key distinction throughout much of the twentieth century was between, on the one hand, the statist model that emerged from the Russian Revolution in 1917 and was extended after the Second World War to other states in eastern Europe, as well as to parts of Asia (China, North Korea, Vietnam) and, on the other hand, the free-market capitalist model that dominated elsewhere. Yet, even that binary distinction was not as neat as it seemed: within the Soviet bloc, countries such as Yugoslavia and China broke away and developed their own characteristics. Within the capitalist world, the free market was also held in check by state regulation and, in many parts of the developing world, by an authoritarian state. By the 1980s two very distinctive models were identified within the so-called free-market capitalist bloc as having led countries to developmental success: the social democratic state, most successful in the Nordic countries of Scandinavia, and the developmental state of Japan and a number of East Asian countries.

While very different from one another, what is common to the social democratic and the developmental state models is the role of the state in regulating and guiding the market for developmental social ends. This was far from the statist model of Eastern European communism which tended to stifle and even eradicate all vestiges of a free market; in Scandinavia and East Asia, the state worked to help the market sector of the economy thrive. The main difference lay in the role of the state in guiding the market: social democracy focused on co-ordinating capital and labour (employers and trade unions) in collaborative agreements that served the interests of each. Such polities were usually characterised by high taxes which funded high-quality and often universal social services. In Japan, South Korea, Taiwan and Hong Kong on the other hand, the state worked hand in glove with private entrepreneurs to develop innovative technological sectors preparing and cajoling them to become competitive in the international market place (Amsden 2001; Woo-Cumings 1999). The first was more a welfare model, whereas the second was very much an economic developmental model in which welfare took a very secondary role. However, in terms of fulfilling their own objectives, the success of each of them rested very much on the proactive role of the state in collaboration with private capital (though state enterprise also played a role in both models).

Meanwhile, the implementation of a radical free-market model in Chile following the military coup of September 1973 by a group of young economists educated in the University of Chicago known as the 'Chicago boys', ushered in a new model that came to be labelled 'neoliberalism' by its critics (Harvey 2005; Green 1995). This was based on strict neo-classical tenets in seeking to free markets from state and other forms of non-market interference, portraying them as neutral areas of exchange that efficiently allocate resources in response to supply and demand signals. This can sometimes be portrayed therefore as a model that relegates the state to a very subservient position with its role being to ensure that markets operate competitively and avoid monopolies emerging to distort competition. Yet, as Selwyn reminds us 'behind the ideology, neoliberal policy relies heavily upon states to reshape class relations in favour of capital, in particular finance capital'. Furthermore, in using the state to discipline society, neoliberal models 'reproduce the subordination of the greater part of society (labourers) to the minority (owners of capital)' (Selwyn 2014: 2). The advent and extension of neoliberalism, therefore, radically refashioned the combination of state, market and society, posing a major challenge to both the social democratic and the developmental state models.

In the early 1990s, debates emerged focusing on different varieties of capitalism, particularly in Europe. Albert (1993) identified two models of capitalism based more on cultural traits, each seen to be antagonistic to the other. One was the Anglo-Saxon free-market capitalism embodied par excellence in the US and the UK following the liberalisation of their economies in the 1980s, and this was seen to characterise the Anglophone world in particular. It was identified as a shareholder model in which companies were run to maximise profits for shareholders. What distinguished the Rhineland model of Germany, the Netherlands, Switzerland and, to an extent, France was its identification of a much broader range of stakeholders, including workers, consumers, citizens and the state as having a stake in companies' success. As Crouch characterises this model: 'The essential idea is a capacity to make long-term decisions that maximize certain collective rather than individual goods' (Crouch 2006: 13). By the early 2000s, the debate had moved on to varieties of capitalism following the distinction made by Hall and Soskice (2001) between liberal market economies (LMEs) and coordinated market economies (CMEs). The former they saw as relying principally on market institutions to organise their relationship with suppliers and workers and to secure finance; in the latter these relationships were mediated by non-market institutions such as trade unions and business organisations. The significance for society of this analysis is best seen in outcomes. As Schneider and Soskice put it: 'Coordinated capitalism reinforces consensus-based political systems in producing egalitarian outcomes and a strong welfare state, and vice versa' (2009: 25). 'In all the liberal market economies, the period since the 1970s has seen the development of flexible labour markets, the substantial dismantling of collective bargaining and effective unionization, the ending of most private sector-driven apprenticeship systems and a massive, largely middle-class expansion of staying-on rates in secondary and increasingly higher education' (ibid.: 28–29). The end result has been to entrench deepening inequality, both in outcomes and in life chances as influenced by education levels. Therefore, different models matter profoundly to social outcomes. This is well captured in debates on measuring development (Box 4.2): as we develop political economy models for the transition to a post-carbon society, robust indicators are going to play a crucial role in ensuring that societies are moving in the right direction.

Box 4.2: Measuring Development: Going in the Right Direction?
A major issue that emerged in the early decades of development was the realisation that economic growth did not necessarily make people's lives better. By the 1960s much research was undertaken on the topic of inequality, showing how the benefits of growth often simply made the rich richer and failed substantially to reduce poverty.

Yet growth in a country's gross domestic product (GDP), namely the extent to which the value of all traded goods and services increases from one year to the next, has continued to dominate economic policy and political interest. From the early 1990s, the United Nations Development Programme (UNDP) sought to broaden measures of development by introducing the Human Development Index which combines measures of economic growth with measures related to education and levels of health (see UNDP 2007).

As addressing climate change becomes ever more urgent, so does the need to include environmental considerations in how we measure successful development. One contribution has come from the UK's New Economics Foundation (NEF) which developed a measure based on five indicators:

- *good jobs*: well-paid jobs with protection for employee rights
- *wellbeing*: people's perception of life satisfaction
- *environment*: levels of carbon emissions
- *fairness*: levels of income inequality
- *health*: avoidable deaths

As Wallis puts it, GDP is like a speedometer: 'it tells you whether your economy is going faster or slower' but it 'doesn't tell you whether or not you are going in the right direction'. As we begin to transition to a post-carbon society, we badly need measures that tell us if we are on the right path. Yet, as Wallis says, while the problems with the current measures are recognised 'we still lack a compelling, coherent, simple alternative narrative' (Wallis 2016).

A focus on social outcomes acts as an important balance on what can often be a predominant focus on how the state and the market interact in constituting political economy models. For example, one distinction between the Nordic social democratic model and the East Asian developmental state model is that the former emerged from the political struggles of the 'red-green' alliance between urban and rural workers to build strong social democratic parties (Senghaas 1985); in East Asia by contrast, the model emerged in authoritarian states with little citizen engagement and a lot of repression. A citizens' movement for democratisation emerged only after the economic transformation had taken off. The example of

China shows us that such a movement is by no means certain of success if the state sees it as a threat to economic advance. The case of Latin America is also instructive in that a wave of 'new left' governments emerged in the region in the early 2000s challenging, at least in rhetoric, the dominance of the neoliberal model and the subservient role played by the state and the citizen within it. Debates rage about the extent to which the new left governments constituted a new 'post-neoliberal' model (Wylde 2012; Macdonald and Ruckert 2009; Burdick et al. 2009), but the role of citizens' movements in the emergence of the phenomenon of the new left is recognised. In other words, this was a new model being created from the bottom-up, challenging the economic and political elites and putting new political forces in power (Silva 2009). This was but one expression of the growing recognition since the 1990s of the role of civil society in constituting development models (Chandhoke 1995). Models, therefore, emerge in various ways but the role of social forces in constituting them needs to be acknowledged.

What, then, has been learnt from this half-century of development efforts that might inform pathways to a post-carbon society? The following attempts to distil some relevant lessons:

1. *There is no single pathway*: Different pathways emerge based on top-down prescriptions but also on bottom-up responses from marginal states and societies, and from social mobilisation from below. Such creative responses, both from states and from society, are still awaited to map out alternative pathways to a low-carbon society.
2. *It is a socio-political project*: While technologies played a major role in pathways to development (e.g., dams or the green revolution), in essence it was socio-political. There is no reason to believe that the pathways to a low-carbon society can, or should be, any less political.
3. *Vested interests undermined more widespread success*: Development fell victim to unequal power relations both at a global level and within most countries. Those that succeeded best were those that 'governed the market', to use Wade's phrase about East Asia (Wade 1990).
4. *Reality varied greatly from rhetoric*: The great universal schemes of political leaders and technocrats positing a single way to development proved illusory. Success was due more to creative responses emerging from within society to pragmatic situations rather than to grand schemes of socio-economic engineering.

5. *Pathways are formed within prevailing ideological parameters*: Any conceptions of socio-economic change must draw from the toolkit of political ideologies which itself is constantly evolving and changing in response to the lessons of lived experiences. These are the only traditions we have that can help shape pathways of radical social change.
6. *New dynamics and human agency*: Recognising the importance of 'historicity' and the legacy of past developments that create structure but also inertia, it is important to consider human agency in future change. New dynamics will unfold and new social and political configurations will evolve. Society, state and market do indeed drive change and new collective visions can motivate desirable change. While it may be difficult in the short-term, change is not only possible but inevitable in the long term (this is the realm of 'futures-thinking' and scenarios, see Chap. 5).

Towards a Political Economy of Climate Change

Applying the lessons from international development experience, therefore, requires focusing much more centrally than is now being done on the political economy configurations that can help us transition to a low-carbon society. This urgently needs to become a much more dominant framework through which to analyse the challenges we face in addressing climate change, placing these challenges in the context of the wider political visions and social struggles that are shaping and reshaping our society worldwide. Yet, the reality at the moment, as evidenced in much transition studies, is that pathways towards a low-carbon society exist without rigorous critique of the wider forces reshaping our society. This 'neutrality' of the analytical project, with insufficient attention to the social, cultural and institutional factors, also needs to be married to a politics of transition, for a holistic vision of what can change and what indeed would be desirable change. This is particularly relevant in the context of the political turbulence emerging in the industrialised countries of the early twenty-first century. The citizenry is increasingly questioning, how and indeed if, the economy and politics are serving its interests. Yet, of course, these factors constitute the context within which we are trying to address the wicked problem of climate change. As Strachan and Foxon put it in discussing their scenarios for low-carbon energy futures, the transition can be based on 'Market Rules,' on 'Central Co-ordination' or on 'Thousand Flowers' scenarios; each of these offer very different pathways (Strachan and Foxon 2012: 86–88). What applies to low-carbon energy applies more widely to a low-carbon society and economy: will it be built primarily on the rules of the market, on the central co-ordination of the state or on the creative endeavours of the countless groups and individuals engaged in a myriad of different forms of civil society activism? Or, perhaps more realistically, on what mix of these elements will it be built?

It is not surprising that, within the hyperfinancialised form of neoliberalism that has dominated the global political economy for the past two decades, the dominant attempts to reduce GHG emissions have entailed using the rules of the market to create incentives for producers and consumers to move to activities that emit less carbon. This has been described as climate capitalism, 'a model which squares capitalism's need for continual economic growth with substantial shifts away from carbon-based industrial development' (Newell and Paterson 2010: 1). Essentially, this entails creating different types of carbon markets which put a price on carbon, such as emissions trading like the EU's Emissions Trading Scheme (ETS) and the Clean Development Mechanism (CDM) which allows investment in developing countries to offset carbon emissions, as well as expanding markets for renewable energy technologies which offer opportunities for investors (see Newell and Paterson 2010 for a comprehensive discussion, and Chap. 8 of this book). The emergence of these market-based mechanisms has helped shift the view of some sectors of business and finance from seeing climate change as a threat to seeing it as an opportunity to be embraced. An authoritative example of this view came in the report of the Global Commission on the Economy and Climate, which as well as some former presidents and prime ministers, contains a number of high-level bankers and entrepreneurs. Entitled *Better Growth, Better Climate: The New Climate Economy Report*, it argues that the next 15 years offer a critical opportunity to undertake 'a deep structural transformation' of the global economy, 'to build lasting economic growth at the same time as reducing the immense risks of climate change. … Future economic growth does not have to copy the high-carbon, unevenly distributed model of the past' (GCEC 2014: 8). Instead 'there is now huge potential to invest in greater efficiency, structural transformation and technological change in three key systems of the economy', cities, land use and energy (ibid.). There is evidence therefore that some of the innovative financial and technological potentials of capitalism are being harnessed to help society make the transition to a low-carbon society. And this has been achieved not just through the endeavours of market actors such as corporations and pensions funds but has also involved governments which have created many of the rules and incentives for these activities, and civil society actors such as NGOs which have criticised the outcomes of the CDM and sought to encourage better standards.[3] It therefore constitutes a political economy model, but one driven by market interests.

However, many questions remain about just how far climate capitalism can take us towards a low-carbon society. Newell and Paterson raise questions about the extent of the benefits it is delivering, highlighting difficulties of methodology in assessing these benefits (such as accounting, measurement and regulation problems) and weak governance of these markets (targets that are not robust enough and rules that are too loose). Based on their analysis of the development of climate capitalism up to now, they outline four possible future scenarios: a climate capitalist utopia where all the mechanisms work to achieve the goal of a low and post-carbon society; stagnation where markets falter and fail to achieve their potential; a 'decarbonised dystopia' which manages to achieve the objective sought but in a highly inegalitarian manner benefiting the privileged and placing the burden of adjustment on the poor and vulnerable, and climate Keynesianism where much stronger governance ensures markets act to achieve the objective sought. All are plausible, they write, but add: 'In all likelihood some messy mix of them all will co-exist—some areas of the world stagnating, others going ahead with a pure neoliberal version, while others still regulate the carbon economy more stringently' (Newell and Paterson 2010: 178).

Newell and Paterson's analysis highlights that, without stronger governance, the outcomes of market-led approaches are likely to result in highly unequal and even dystopian futures, as illustrated in Box 4.3. This focuses attention on the role of the state. Drawing explicitly on the lessons of the developmental states of East Asia, Bailey and Preston argue that a low-carbon transition 'is highly interventionist in nature' requiring states with the 'capability to develop, through inclusive processes, a national transformation project and translate this into discrete plans, targets and policy measures' (Bailey and Preston 2014: 2, 3).

Duit et al. refer to this as the 'environmental state' and they identify four dimensions: as a system of regulation with laws, rules, controls, special facilities and protections all directed to environmental concerns; as an administrative apparatus including environmental ministries and specialist agencies, as well as assessment and advisory bodies; as a corpus of ideas and expert knowledge seeking to promote greener values and attitudes and gain legitimation for state action on their behalf; and as an arena for environmental conflict and a site for authoritative decision-making with lobbying, consultation and negotiation around defining and deciding on environmental issues (Duit et al. 2016: 7–8). While the focus here is on the national state, it is also recognised that the scale of social transformation required will entail changes at supranational as well as at subnational levels. For example, the German Advisory Council on Global Change (WBGU) advocates the mainstreaming of climate change policy at EU level; while this already happens, it 'needs a stronger commitment to transformative policies across all organisational units' (WBGU 2011: 219). The environmental state therefore constitutes a second political economy model, this time centred on a strong and transformative state.

However, such a state needs to balance two principles that have for long been seen as being in opposition: on the one hand, empowering the state to actively determine priorities (a task which, over recent decades, has been handed over to private market actors) and, on the other, 'providing citizens with more extensive opportunities to have a voice, to get more involved in decision-making processes, and to take on a more active role in politics' (ibid.: 209). This is what Duit et al. call 'ecological citizenship' possibly leading to a 'Green social contract' making explicit the environmental rights and obligations of the state and its citizens (Duit et al. 2016: 12). This then moves into a third political economy model, one in which citizens are empowered to hold the state to account as it leads society towards a low-carbon future. But the role of the citizen goes beyond being a check on the power of the state and becomes an essential driver of ecological innovation (hence Strachan and Foxon's term of Thousand Flowers for this third model). We can call this third model, the ecological innovation model. While the term innovation is regularly invoked in regard to science and technology, the transition to a low-carbon society is also dependent on a lot of social innovation, experimenting with new ways to live within our ecological footprint. In a paper for the European Union on transitional governance in the service of sustainable societies, Belgian political scientist Olivier De Schutter emphasises the 'role of social innovations empowering people to invent local solutions' (De Schutter 2014: 17). He writes that these social innovations abound and they are often local and territory-based. 'They typically are based on hybrid governance structures, bringing together municipalities, the private sector, the "third sector" and non-governmental organisations or citizens' groups' (ibid.). He gives the example of transition towns 'in which neighbours work together towards improving energy efficiency, community building and domestic micro-generation installation' (ibid.). Another example he mentions is Cloughjordan Ecovillage in Co. Tipperary, Ireland 'a supportive social community living in a low-impact way to create a fresh blueprint for modern sustainable living' (ibid.) (see Kirby 2017).

De Schutter emphasises the vital importance of innovative experiments since 'they expand the range of alternatives people may choose from. What emerges is a participatory society that includes, but goes beyond, both participatory democracy and consumer activism' (ibid.: 29). Instead of being beneficiaries of public policies, clients of business, or stakeholders in society, citizens become 'actors' or 'searchers' making an active contribution to designing solutions, implementing them in various contexts and assessing their contribution to the transition to a low-carbon future. In terms of governance, local levels of decision making are a crucial asset in designing solutions that represent true alternatives: 'the number of veto points is smaller, and the possibilities for synergies across sectoral policies (such as energy, mobility, food, and education) are greater' (ibid.: 27). If, as the WBGU does, we understand the transition to a low-carbon society as something similar to the transition from pre-industrial to industrial society, then the scale of the transformation required in a short space of time is breathtaking. For this reason, an unleashing of social creativity is required to contribute to the wider society the lessons of what can work in making this transition successfully and in a just way. It will involve what De Schutter calls 'a high-intensity politics ... as the distinction breaks down between the roles individuals occupy as consumers and producers, as members of their families and their communities, and as citizens (ibid.: 27).

Box 4.3: Governing Global Energy Markets: Mind the Gaps

Global energy governance (GEG) has emerged as a field of study driven by three factors that are transforming the supply of energy: climate change, geopolitical tensions and increasing volatility in supply. Since these impact on economic performance and human well-being throughout the world, understanding how they are governed is of major public importance.

As Van de Graaf and Colgan state 'market price signals alone are often insufficient to provide satisfactory outcomes', so that there is a need for some form of governance to deal with possible energy shortages at national and international level, nuclear proliferation and climate change (Van de Graaf and Colgan 2016: 3). Yet, they identify notable gaps between 'the actual and potential scope of GEG': a patchwork of institutions exists—intergovernmental, multilateral financial institutions, international NGOs and hybrid entities—but 'there is no single core to the complex' due to multiple and conflicting interests between the many players involved (ibid.: 8).

Two practical consequences are given. The first relates to international security as oil is a factor in both civil wars in oil-producing countries and in international conflicts. Yet, 'the regime complex for energy has no real institutional capacity for addressing these issues'. The second relates to the protection of human rights in the developing world as there is no system for tracing how oil is fuelling human rights abuses such as those committed by Islamic State in Syria (ibid.: 7).

Different political economy models to address climate change and the transition to a low-carbon society do, therefore, already exist, at least in embryo. The dominant one is climate capitalism, but elements of an environmental state model are also identifiable and, as De Schutter's examples illustrate, so too are elements of an ecological innovation model. Yet, as Bailey and Preston state, the many initiatives being undertaken through today's dominant power configurations 'are not translating to emissions reductions at anything like the pace needed' (Bailey and Preston 2014: 2). This raises urgent questions about developing models that can more effectively lead to a low-carbon society. As Meadowcroft put it in discussing governance for sustainable development:

> It is not something that competent officials can get done quietly and efficiently, out of the public view, as citizens go about their everyday business. On the contrary, it is an inherently political process—for it requires societal decisions about desirable ways of life, and about how benefits and burdens are to be shared among different communities and different generations, and between humankind and other inhabitants of this planet. The sort of radical decoupling of economic activity from environmental burdens that sustainable development implies will require iterative processes of reform stretching over many decades. (Meadowcroft 2010: 313)

Conclusions

This chapter has examined and given substance to issues raised by the IPCC as being crucial to the transition to a low-carbon society but which are treated in a cursory way in the literature of transition studies. These include issues of governance, new institutions and institutional arrangements, ethical dimensions and sustainable development. Recognising that the project of international development since the Second World War is the nearest experience we have to draw on from which to learn lessons which can assist in moving countries on pathways of social and economic transformation, the chapter has examined the lessons we can learn from half a century of development efforts. It has sought to address the issues raised by Robinson's conclusion that 'sustainable development approaches should move beyond a technocratic, nature-centred view and explicitly address issues of power, the distribution of wealth, and the local grounded experience of natural and human-made processes of production' (Robinson 2004: 370).

Other authors have mapped out some similar terrain and identified different approaches that inform pathways towards a low-carbon society. For example, in their discussion of 'the political economy of the global environment', Clapp and Dauvergne identify four environmental worldviews or visions: the market liberal vision, the institutionalist vision, the bioenvironmentalist vision and the social green vision (Clapp and Dauvergne 2011). Similarly Bäckstrand and Lövbrand identify three contending climate governance discourses that inform current debates in the post-Copenhagen era: ecological modernisation, green governmentality and civic environmentalism (Bäckstrand and Lövbrand 2016). With the exception of Clapp and Dauvergne's bioenvironmentalist vision, each of these map closely to the three political economy models of climate change identified in this chapter. But it has moved beyond seeing them as different discourses or visions and identified them as contending political economy models each of which configures the state-market-society relationship in very different ways. It thus brings power configurations to the centre of the discussion. This discussion will be further developed in later chapters, particularly Chaps. 8 and 9 that respectively examine the dominant climate capitalism model and the potential for an ecosocialist model. In the next chapter, we examine how scenarios can allow us to explore and envision future change. To 'think outside the box' and generate new dynamics and agency beyond current limitations, as we seek to transition to post-carbon and sustainable societies.

Notes

1. The IPCC seeks to be policy relevant but not policy prescriptive as part of its remit. Coming to a more critical examination of transformation pathways requires a stronger focus on the constitution of development pathways and the politics of what they entail.
2. This is to be distinguished from the 'third way' politics that emerged in Western countries in the 1980s. This was essentially a centrist approach that married right-wing economic orthodoxy to left-wing social policies (Bobbio and Cameron 1997) as a modified form of neoliberalism.
3. The 'Gold Standard' methodology to certify CDM projects was devised by the World Wildlife Fund and other NGOs using more strict criteria. In response to concerns of unsustainable projects and spurious credits, this included only allowing renewable energy projects.

References

Adelman, Irma, and Cynthia Taft Morris. 1973. *Economic Growth and Social Equity in Developing Countries.* Stanford, CT: Stanford University Press.

Albert, Michel. 1993. *Capitalism Against Capitalism.* London: Whurr.

Alkire, Sabina. 2010. Human Development: Definitions, Critiques, and Related Concepts. *Working Paper No. 36*, Oxford Poverty and Human Development Initiative (OPHI). Oxford: OPHI.

Amsden, Alice H. 2001. *The Rise of "The Rest": Challenges to the West from Late-Industrializing Economies.* Oxford: Oxford University Press.

Bäckstrand, Karin, and Eva Lövbrand. 2016. The Road to Paris: Contending Climate Governance Discourses in the Post-Copenhagen Era. *Journal of Environmental Policy & Planning*, published online March 2016, available at http://www.tandfonline.com/doi/abs/10.1080/1523908X.2016.1150777

Bailey, Rob, and Felix Preston. 2014. Stuck in Transition: Managing the Political Economy of Low-carbon Development. *Chatham House Briefing Paper*, February 2014.

Baker, Susan. 2006. *Sustainable Development.* London: Routledge.

Bobbio, N., and A. Cameron. 1997. *Left and Right: The Significance of a Political Distinction.* Chicago: University of Chicago Press.

Burdick, John, Philip Oxhorn, and Kenneth M. Roberts, eds. 2009. *Beyond Neoliberalism in Latin America? Societies and Politics at the Crossroads.* Basingstoke: Palgrave Macmillan.

Cardoso, Fernando Henrique, and Enzo Faletto. 1979. *Dependency and Development in Latin America.* Berkeley: University of California Press.

Chandhoke, Neera. 1995. *State and Civil Society: Explorations in Political Theory.* New Delhi: Sage.

Clapp, Jennifer, and Peter Dauvergne. 2011. *Paths to a Green World: The Political Economy of the Global Environment*. Cambridge, MA: The MIT Press.

Crouch, Colin. 2006. Models of Capitalism. In *Key Debates in New Political Economy*, ed. Anthony Payne, 11–31. London: Routledge.

De Schutter, Olivier. 2014. *The EU's Fifth Project: Transitional Governance in the Service of Sustainable Societies*. Accessed 19 February 2015. http://www.srfood.org/images/stories/pdf/otherdocuments/Framing4.pdf

Di Muzio, Tim, and Richard H. Robbins. 2016. *Debt as Power*. Manchester: Manchester University Press.

Duit, Andreas, Peter H. Feindt, and James Meadowcroft. 2016. Greening Leviathan: The Rise of the Environmental State? *Environmental Politics* 25 (1): 1–23.

Global Commission on the Economy and Climate. 2014. *Better Growth, Better Climate: The New Climate Economy Report*. Washington, DC: New Climate Economy.

Green, Duncan. 1995. *Silent Revolution: The Rise of Market Economics in Latin America*. London: Cassell.

Hall, Peter, and David Soskice. 2001. *Varieties of Capitalism: The Institutional Foundations of Comparative Advantage*. Oxford: Oxford University Press.

Harvey, David. 2005. *A Brief History of Neoliberalism*. Oxford: Oxford University Press.

Hayter, Teresa. 2005. Secret Diplomacy Uncovered: Research on the World Bank in the 1960s and 1980s. In *A Radical History of Development Studies: Individuals, Institutions and Ideologies*, ed. Uma Kothari, 88–108. London: Zed Books.

Hoogvelt, Ankie. 1997. *Globalisation and the Postcolonial World: The New Political Economy of Development*. Basingstoke: Macmillan.

IPCC. 2014. *Climate Change 2014: Synthesis Report*. Geneva: IPCC.

Kay, Cristóbal. 1989. *Latin American Theories of Development and Underdevelopment*. London: Routledge.

Kiely, Ray. 2007. *The New Political Economy of Development: Globalization, Imperialism, Hegemony*. Basingstoke: Palgrave Macmillan.

Kirby, Peadar. 1997. *Poverty Amid Plenty: World and Irish Development Reconsidered*. Dublin: Trócaire and Gill & Macmillan.

———. 2017. Cloughjordan Ecovillage: Modelling the Transition to a Low-Carbon Society. In *Transitioning to a Post-Carbon Society: Degrowth, Austerity and Wellbeing*, ed. Ernest Garcia, Mercedes Martinez-Iglesias, and Peadar Kirby, 183–205. Basingstoke: Palgrave Macmillan.

Lélé, Sharachchandra M. 1991. Sustainable Development: A Critical Review. *World Development* 19 (6): 607–621.

Macdonald, Laura, and Arne Ruckert, eds. 2009. *Post-Neoliberalism in the Americas*. Basingstoke: Palgrave Macmillan.

Meadowcroft, James. 2010. Who is in Charge Here? Governance for Sustainable Development in a Complex World. *Journal of Environmental Policy & Planning* 9 (3–4): 299–314.

Munck, Ronaldo. 2013. *Rethinking Latin America: Development, Hegemony, and Social Transformation*. Basingstoke: Palgrave Macmillan.

Nederveen Pieterse, Jan. 2001. *Development Theory: Deconstructions/ Reconstructions*. London: Sage.

Newell, Peter, and Matthew Paterson. 2010. *Climate Capitalism: Global Warming and the Transformation of the Global Economy*. Cambridge: Cambridge University Press.

Payne, Anthony. 2005. *The Global Politics of Unequal Development*. Basingstoke: Palgrave Macmillan.

Revi, Aromar. 2014. Opinion: IPCC Adaptation Report Points the Way to Climate Resilient Development Pathways. Climate and Development Knowledge Network (CDKN Global). Accessed 11 May 2016. http://cdkn.org/2014/04/opinion-enabling-regional-transitions-to-climate-resilient-development-pathways/?loclang=en_gb

Rist, Gilbert. 1997. *The History of Development: From Western Origins to Global Faith*. London: Zed Books.

Robertson, Margaret. 2014. *Sustainability: Principles and Practice*. London: Routledge.

Robinson, John. 2004. Squaring the Circle? Some Thoughts on the Idea of Sustainable Development. *Ecological Economics* 48: 369–384.

Schneider, Ben Ross, and David Soskice. 2009. Inequality in Developed Countries and Latin America: Coordinated, Liberal and Hierarchical Systems. *Economy and Society* 38 (1): 17–52.

Selwyn, Ben. 2014. *The Global Development Crisis*. Cambridge: Polity Press.

Senghaas, Dieter. 1985. *The European Experience: A Historical Critique of Development Theory*. Leamington Spa; Dover: Berg.

Silva, Eduardo. 2009. *Challenging Neoliberalism in Latin America*. Cambridge: Cambridge University Press.

Stewart, Frances. 1985. *Planning to Meet Basic Needs*. Basingstoke: Macmillan.

Strachan, Neil, and Timothy J. Foxon. 2012. A Low-Carbon Transition. In *Living in a Low-Carbon Society in 2050*, ed. Horace Herring, 75–91. Basingstoke: Palgrave Macmillan.

UNDP. 2007. *Human Development Report: Fighting Climate Change*. New York: Palgrave Macmillan.

Van de Graaf, Thijs, and Jeff Colgan. 2016. Global Energy Governance: A Review and Research Agenda. *Palgrave Communications*, January 2016. Accessed 25 May 2016. www.palgrave-journals.com/palcomms

Wade, Robert. 1990. *Governing the Market: Economic Theory and the Role of Government in East Asian Industrialization*. Princeton: Princeton University Press.

Wallis, Stewart. 2016. Five Measures of Growth that are Better than GDP. Accessed 11 May 2016. www.weforum.org/agenda/2016/04/five-measures-of-growth-that-are-better-than-gdp/

WBGU. 2011. *World in Transition: A Social Contract for Sustainability*. Berlin: WBGU.

WCED. 1987. *Our Common Future*. Oxford: Oxford University Press.

Woo-Cumings, Meredith, ed. 1999. *The Developmental State*. Ithaca: Cornell University Press.

Wylde, Christopher. 2012. *Latin America after Neoliberalism: Developmental Regimes in Post-Crisis States*. Basingstoke: Palgrave Macmillan.

CHAPTER 5

Planning Future Pathways: Implications and Outcomes of Scenario Studies

Introduction

Scenarios are about the future, an undiscovered country where change is not only possible, it is inevitable. The question is, can this be harnessed towards a post-carbon and sustainable society, or will we remain slaves to the limitations of the past? The last chapter discussed the legacy of decades of development studies. This chapter looks at the use of scenarios for insight into the future, how scenarios have come to dominate our approaches to viewing the low-carbon transition, and also how insights from the social sciences could be brought to bear in understanding how human systems can change, and the implications of these changes. We return to the contribution of the social sciences later (see Box 5.3). In order to facilitate greater understanding of contribution of the scenario approach, the first section reviews different techniques that are used to explore and analyse the future. We then discuss environmental and emissions scenarios, and what they tell us. This is followed by a discussion of the practice of transition and transformation, and a review of some of the key outcomes from such studies, giving some examples of influential scenarios. Transition and transformation are then critically appraised, arguing that what is needed is a more fundamental transformation beyond the limited reliance on techno-economic measures, to look at sustainability and social, cultural and political drivers. As a development approach

P. Kirby, T. O'Mahony, *The Political Economy of the Low-Carbon Transition*, International Political Economy Series,
DOI 10.1007/978-3-319-62554-6_5

to mitigation, this would not only facilitate a faster and more complete reduction in emissions to give a greater chance of meeting climate goals, but also the achievement of development win-wins. The political economy approach is highlighted as a lens on the many variables of power and politics that are central to the prospects for a social transformation on the scale that is necessary.

Approaches for Insight into the Future

During the second half of the twentieth century, methods used to develop insights into the future moved from the fringes of public policy towards the centre as the challenges policy needed to address became more complex. Rather than blindly stumbling into the future without preparation, or indeed using guess-work and gazing at a crystal ball, the approaches sought to become more systematic and 'scientific' using the evidence provided by studies of the future. Economic forecasting and population projections were early examples of this move, which could enable better public policy planning, the prediction of infrastructure needs and the demands on public services and how to facilitate economic development. Scenario planning emerged from a strategic branch of looking into the future in Hermann Kahn's US military planning of the 1950s, and Gaston Berger's '*la prospective*' process to develop scenarios of the future as a guide in formulating public policy. The oil crisis of 1973 was an important juncture, as the impact of this event pushed analysts to consider how such surprises could occur. Scenario planning grew significantly through the 1970s as a corporate strategy tool pioneered by companies such as Shell, driven at least in part by these events. From 1970s onwards scenarios grew further as a scientific tool for analysis of environmental problems, energy and climate change. The issues involved are complex and long-range, and forecasting and projection techniques were proving inadequate to the tasks.

While forecasts and projections may be useful in some circumstances, they come with many caveats and limitations, particularly through using historical statistical relationships to look into the future. These relationships can change with profound consequences. The constant revision of economic forecasts and population projections, and the failure to foresee the financial and economic crisis of 2008, are illustrations of the limitations of these approaches. The problem with forecasts in the lead up to the crisis, arguably the worst since the Great Depression of the 1930s, is that fundamental problems were not foreseen by most forecasters, apart from a few analysts such as Nouriel Roubini and Robert Shiller. They described potential bubbles rising in housing and stock markets, but were roundly dismissed. Roubini was dubbed 'Dr. Doom' by the media while the 'official future' of unabated global growth continued unchallenged.

It has become clear that the future is not a continuation or repetition of the past. Change is indeed inevitable, but what kinds of change occur can be influenced by collective will. Sometimes known as 'human agency', this view acknowledges that there are 'structural' influences that can cause or obstruct change, but we still have considerable ability to change course. While achieving change is difficult in the short term, in the longer term there is great scope to change our path. Our societies, our politics and our individual decisions dictate how the future unfolds. Consequently developing insights into what changes could occur, or indeed what changes we desire, are valuable exercises when faced with such great societal challenges as low-carbon transition, growing social inequality or the need to improve human wellbeing and environmental sustainability.

The future is characterised by uncertainty and confounds predictive and forecasting approaches that deal with single and 'probable' futures. Scenario analysis and scenario planning, or 'scenario thinking', are approaches used to analyse and prepare for the future, by the conscious consideration of several possible alternatives. This is aided by considering not just quantitative variables, such as the economy and technology, but qualitative social, cultural and political influences. Sometimes termed 'future worlds', this approach is linked to disciplines of study known as 'foresight' and 'futures studies'. They do not look just at outcomes, but at the paths that lead to their evolution. In more general terms, an important thinker in scenario circles, Hugues de Jouvenel, proposed that computer simulation models[1] based on observations of the past are favoured by economists, econometrists, statisticians and forecasters (de Jouvenel 2000: 45). The accuracy or scientific quality of forecasts is not guaranteed and results can be subjective as an arbitrary collection of quantitative variables.[2] De Jouvenel writes that this quantitative approach 'has long been opposed to the scenario method, which is more developed and used by futurists for one simple reason: better a rough but fair estimate than a refined yet incorrect forecast' (de Jouvenel 2000: 45). Rather than putting all the eggs in one basket, scenarios can be used to look at a number of alternative outcomes and help to understand and prepare for the risks and opportunities that the future holds.

Scenario analysis has a rich history spanning decades in an increasing number of sectors and disciplines. As discussed by O'Mahony (2014: 42–43), scenarios are a tool to deal with the uncertain future, but they also have a range of other uses. An expanding array of business, community, policy and research contexts use scenarios with highly varied objectives—better management, consciousness raising, conflict resolution, policy advice and research (Raskin et al. 2005: 36). Scenario analysis has become particularly important in understanding the challenge of climate change, sustainability and how to 'transition' to a low-carbon future.

Environmental Scenarios and What They Tell Us

As environmental and sustainability issues are long-term challenges, scenario analysis has been used to explore what our current patterns of development mean for future human and environmental wellbeing. Environmental issues are not only important for the potential damages inflicted on the natural world, on ecosystems, habitats and biodiversity but on 'environmental media' such as air, water and soil. These provide 'ecosystem services' on which human existence relies. There are profound implications for society, the environment and our legacy to future generations, as current human activities are placing them in peril. We need tools to analyse what can plausibly change in the future as technical scientific exercises, dubbed 'inquiry-driven' scenarios by Alcamo et al. (2008: 8), but we also need tools for imagining, discussing and preparing for the future. These latter scenarios are essentially social and political tools that help us to strategise our responses, as visions of a sustainable future and the paths needed to achieve them.

The pressing nature of environmental challenges has driven comprehensive scenario exercises to raise awareness of current global problems, and what needs to be done to address them. The Global Environmental Outlooks (GEO) are periodic reports issued by the United Nations Environment Programme (UNEP). The currently ongoing UNEP GEO 6 process (with the final report due by 2018) has already documented increasing rates of environmental damage across the planet, but with potential to reverse these if policy is urgently implemented. The report places particular focus on consumption and production as drivers of environmental damage (UNEP 2016). It points to low-carbon, climate-resilient choices in infrastructure, energy and food production, coupled with effective and sustainable natural resource governance. A gradual steady transition towards sustainability and the opportunities it presents is recommended, while poverty eradication remains a central goal. The UNEP GEO 5 report entitled; '*Environment for the future we want*' placed emphasis on including technological responses, but the report also recommended moving beyond the focus on technology (UNEP 2012). The report highlights the need to shift the policy focus to address 'underlying drivers' such as consumption patterns.

Consumption is a recurring issue in climate change and sustainability policy (Fleurbaey et al. 2014). Effective policy to address 'over-consumption', predominantly in developed countries, has been sorely lacking, with many negative effects of this growth in consumption levels and its global spread. Consumption is a critical issue of global social inequality, and as such, is an issue of power dynamics for political economy. Box 5.1 discusses some of the issues involved.

Box 5.1: The Place of Consumption and Wellbeing in the Low-Carbon Transition

Debates on the place of consumption in human wellbeing have a philosophical tradition that spans millennia. This thorny question has become more pressing since the early 1990s when global sustainable development and climate change treaties were being debated. A key ethical issue is that consumption currently facilitates the luxury and wellbeing of the wealthy at the expense of the poor, while also driving environmental damage and GHG emissions. It is also recognised that consumption has many damaging effects on individual and societal wellbeing of the wealthy, and that these patterns are spreading to developing countries (Fleurbaey et al. 2014: 308). This is a 'lose-lose' situation as various related environmental problems, including attempts to stall climate change, are becoming more intractable as consumption continues to grow.

But if there are negative effects of consumption on the individual, society and the environment then how do we change course? The field of Sustainable Consumption and Production (SCP) has focused on efficiency on the production side and behaviour on the consumption side with limited success. Researchers are now exploring a more fundamental approach described by Professor Tim Jackson at the University of Surrey as the 'double dividend', enhancing human wellbeing while reducing emissions as a win-win. The MAXWELL project at the Finland Futures Research Centre considers the use of 'sustainable wellbeing', by placing priority on other more beneficial life domains in future scenarios for the EU. The concept seeks to explore how adding to people's lives, rather than loss and sacrifice, could be a fruitful approach to addressing the problem of overconsumption. This is an area that requires significant policy development and greater research efforts.

The Millennium Ecosystem Assessment (MEA) was a major assessment of the human impact on the environment. Called for by then United Nations Secretary-General Kofi Annan in 2000, it was published in 2005. It provides a scientific appraisal of the condition and trends in the world's ecosystems, the services they provide, and options to restore, conserve or enhance the sustainable use of ecosystems. The MEA Scenarios Working Group (Carpenter et al. 2005: 2) considered the possible evolution of ecosystem services during the twenty-first century by developing four global scenarios. They concluded that:

1. significant changes in policies, institutions and practices are required,
2. these can mitigate some but not all of the negative consequences of growing pressures on ecosystems, and
3. the changes required are substantial and are not currently under way.

The MEA scenarios re-affirmed that it is human activities, our patterns of development and the policies we implement that are at issue. These will dictate whether we continue on the current unsustainable path or in a direction that maintains our ecosystems while facilitating human wellbeing through a sustainable form of development. Box 5.2 discusses a historic environmental scenario study that has created much debate since its publication in the 1970s.

Box 5.2: 'The Limits to Growth', an Early Environmental Controversy

One of the earliest sustainability studies, known as 'the limits to growth' (Meadows et al. 1972) was a scenario study commissioned by the Club of Rome to explore the potential global consequences of the interactions between human activities and the Earth's systems. Researchers at Massachusetts Institute of Technology, including husband-and-wife team Donella and Dennis Meadows, built a computer simulation model to track the world's economy and environment known as 'World3'. A set of different world population, industrialisation, pollution, food production and resource depletion assumptions in each future scenario were fed into the World3 model. The results showed that in two of the scenarios, an overshoot and collapse of the global system that supports humanity occurred by the mid-to-late twenty-first century. Another scenario resulted in a 'stabilised world' through applying limits to production of material goods and achieving equilibrium.

The 'Limits to growth' continues to generate debate to this day. Criticism has often focused on the conservative technological improvement assumptions that were used, and also the perceived 'inaccuracy' of the scenario results. A recent review of the scenario results suggests that they are actually 'within uncertainty bounds of nearly all the data in terms of both magnitude and the trends over time' (Turner 2008: 37) and are tracking close to the 'Business As Usual' scenario in the original report. However, while recognising that the scenarios were not intending to achieve 'accurate predictions', there is a far more prescient observation from a policy and awareness perspective. The 'limits to growth' report firmly put the question of the sustainability of global development trends on the public agenda for the first time. This pierced the rigid and steadfast belief that growth was infinite, to the incredulity of many economists contemporary to the era. A number of major global environmental challenges have since manifested in the intervening years, not least human-induced climate change, pointing indubitably towards the now well acknowledged necessity to radically transform the sustainability of global development.

Scenarios have become standard in analysing climate change impacts and policy, looking at environmental pressures such as air pollution and water extraction, examining ecosystem damage, exploring future sustainability and many other domains that might be perceived as 'environmental'. Scenarios are also used prominently by intergovernmental bodies such as the United Nations Environment Programme (UNEP), the International Energy Agency (IEA) and the European Environment Agency (EEA). But the scenarios produced are at their core, at least in a broad sense, actually development scenarios, whether acknowledged or not. It is the impact of human activities that we are seeking to understand and this is by necessity a question of how we develop.

Emission Scenarios, Efforts to Understand Climate Change Expand

As climate change is unequivocally driven by anthropogenic GHG emissions,[3] the evolution of the human systems that lead to these emissions is the core interest of efforts to avoid dangerous anthropogenic interference with the climate.[4] Energy consumption, in particular fossil fuels such as coal, oil, gas and peat, from power generation to industrial production and from how we heat our homes to how we power our vehicles, is the largest source of greenhouse gases and is consequently the starting point for emission scenarios. Future energy consumption is explored in economic sectors by energy models that use approaches to understand the relationship of economic activity to energy, in how much energy is required for each of our activities and the production of goods and services. But the way that energy models are structured according to 'economic and technological realities' is not an entirely objective exercise (Nielsen and Karlsson 2007). Requiring economic theories and assumptions, they can reflect specific futures that are profitable or preferable to certain interests or can be used to legitimise results rather than guide policy (Midttun and Baumgartner 1986). All analysis is therefore intrinsically also a political and ethical exercise which by definition cannot entirely separate science from politics.

The increasing global policy and scientific importance of climate change and environmental challenges, particularly towards the end of the twentieth century, drove an advancement in the science of looking at the future. Scenarios were adopted by the Intergovernmental Panel on

Climate Change (IPCC) as the appropriate tool to explore future global GHG emissions to 2100 and beyond, given the uncertainty of future development and related GHG emissions. The Special Report on Emission Scenarios (SRES) of the IPCC was a watershed in scenario development (Nakicenovic et al. 2000); it explored global development futures and temperature increases associated with each scenario. It was highly influential in the science of climate change as it allowed scientists to consider what the future rates of GHG emissions could be in various scenarios, and what this would mean in terms of physical changes in the climate. This allowed the IPCC to deepen its analysis of the causes of, and solutions to, climate change in its 'Assessment Reports' that advise policymakers throughout the world. As it sought to move beyond forecasts and projections, it also began the process of shifting the focus of attention away from solely economic and technological views, to social, cultural and political driving forces of how the world evolves. This was a valuable contribution to our knowledge but one that remained largely limited to the academic community in the field of climate change science and policy. It is important to note that these emission scenarios were not predictions. They were explorations of different plausible future worlds that would evolve without policy to change course, which becomes the subject of transition and transformation scenarios. As the scenarios showed increasing GHG emissions and related global temperatures without radical action, it achieved its goal of provoking much greater and more informed global policy debate on future climate change and the implications for development.

A dualism in emission scenarios existed in two largely non-overlapping streams of inquiry through quantitative modelling and qualitative narratives (Fisher et al. 2007: 174). The main differences between the model-based and non-model based scenarios is the technical and economic detail in the former and social, cultural and political developments in the latter (Nielsen and Karlsson 2007: 305). Scenarios can be used as linking tools of qualitative narratives about the future and quantitative formulations based on modelling. Box 5.3 reviews the potential contribution of the social sciences to the practice of scenario development and understanding transition. In understanding the potential contribution of development studies, political economy and indeed the social sciences in general, it is useful to discuss the history of scenario practices.

Box 5.3: The Contribution of the Social Sciences to the Practice of Scenarios and the Understanding of Transition

In this book, we concentrate mainly on the current contributions from various strands of *economics*, and the potential contribution of *political economy*. As a branch of the social sciences, political economy can enhance the practice of scenarios and the understanding of transition. Research on scenarios and transition already includes much *economic* thinking and some insights from the fields of *political science* and *demography*. There is much to be gained from greater inclusion of insights in *political economy*, *sociology*, *psychology*, *cultural studies*, *human geography*, *moral philosophy* and *ethics*, to understand both the processes and outcomes of change, and their implications. The humanities could make a compelling contribution through fields such as *history*, *anthropology* and *philosophy* in general, through the understanding of deep questions on how societies change and what this means, offering valuable alternative perspectives. However, as future scenarios are partly a creative process, perspectives and re-interpretations from *the arts* can open up entirely new dimensions which may not be immediately obvious to other disciplines. As the future is complex and uncertain, transdisciplinary approaches to scenarios are considered the gold-standard, and an increasing role for the social sciences in general is necessary to embrace the multitude of driving forces that dictate how the future unfolds.

Trajectories of future energy carbon emissions are determined by complex dynamics as discussed in the IPCC Fourth Assessment Report in Fisher et al. (2007: 178). In Chap. 3 'reducing emissions the Kaya way' we looked at the Kaya identity (Kaya 1990), an approach to separate and simplify the driving forces of energy emissions. The focus of the Kaya identity is essentially population, economic growth and technological change through efficiency and decarbonisation by renewables. But in the context of exploring the driving forces of emissions there are two important framing conditions. Firstly, GHG emissions arise from more than just the combustion of fossil fuels; agricultural emissions such as those from the rearing of livestock are a significant factor, while industrial process emissions such as cement production, and land use change and deforestation are also significant sources. Secondly, there are contextual factors, driving forces that are 'superordinate', that have a strong effect on the change in Kaya driving forces. As discussed in the scenario driving force literature review of O'Mahony et al. (2013), and O'Mahony (2013), these factors could be articulated in various forms, but for simplification, they could be described as social, cultural and political driving forces, which call strongly for the inclusion of different categories of social scientists in scenario development teams.

Emission scenarios have been an important part of the IPCC process since its inception, beginning with the IS92 scenarios which were replaced by the SRES scenarios in 2000. Each SRES storyline represented different demographic, social, economic, technological and environmental development pathways. It did however leave somewhat of a gap in comprehensively looking at social, cultural and political driving forces. A new scenario process was established for the IPCC Fifth Assessment Report. For the research communities involved in mitigation and transformation analysis and policy, a process of developing 'Shared Socioeconomic Pathways' (SSPs) was implemented. The SSPs of O'Neill et al. (2015) describe plausible alternative changes in aspects of society such as demographic, economic, technological, social, governance and environmental factors. There is much to commend these new scenarios with the inclusion of human development, lifestyle and consumption, and also in how they offer strategic insights, including a greater understanding of development and social, cultural and governance drivers. Further research on detailing SSPs could benefit from a more concentrated focus on these drivers and more inclusion of experts from the social sciences.

The question that arises procedurally is how comprehensively such scenarios encompass social sciences and political economy? From a political perspective the question is: what is the place for public participation and what are the social and political implications of the scenarios that are developed? As these emission scenarios are predominantly scientific 'inquiry' exercises, these issues are even more prescient when it comes to strategy and policy-related scenarios in the new field of 'transition' and 'transformation'. While emission scenarios allow us to understand what the problem is, and its consequences, the new field of transition and transformation scenarios seeks to explain how we can move to a post-carbon future, to address the problem and work towards the many opportunities it presents. This is a related but separate task to emission scenarios and it is to these studies that we move next.

Transition and Transformation Scenarios, from Technology to Political Economy

As efforts to reduce emissions are now being intensified globally in the post-Paris Agreement period, the requirement for long-term visions of how to change course and reduce emissions in line with the science will continue to drive the use of long-term scenarios. We have discussed the use of scenarios in general, for 'environmental' and sustainability issues, and more specifically emission scenarios. We now turn to a particular type of scenario that is used to analyse change and catalyse policy, specifically for transition and transformation.

In order to understand such scenarios it is necessary to clarify some of the terms that are used. The analytical emissions scenario literature frequently refers to pathways and sustainable development pathways, while the more strategic and policy-related literature refers to transition and transformation. We referred to 'development pathways' in Chap. 3, and the definition there could be further distilled to the set of driving forces that evolve and interact in determining the direction and characteristics of development over time. 'Sustainable development pathways' are also discussed in Chap. 3, and could be defined as an equitable balance of human development, wellbeing and the environment within a development pathway. With an important role for appropriate policy processes through 'governance',[5] sustainable development pathways are intrinsically linked to transition in that they inherently facilitate lower emissions development paths. Yet, there remains a need for a clarification of the distinction between transition and transformation. The special report on adaptation to climate change, extreme events and disasters of the IPCC (2012: 5) defines 'transformation' as: 'The altering of fundamental attributes of a system (including value systems; regulatory, legislative, or bureaucratic regimes; financial institutions; and technological or biological systems).' Whereas 'transition' tends to refer to the energy system or the economy changing gradually over time towards a 'low-carbon' future, transformation refers to fundamental systemic change. Chappin and Ligtvoet (2014) discuss the differences between transition and transformation, describing transition literature as linked to sustainability as a normative goal and energy as the dominant topic. However, in transformation literature, energy and sustainability are often placed within larger processes of societal change, that are especially relevant to political economy, such as economic development, demographics, or the change from a communist to a capitalist society. From a policy perspective, transformation could be more desirable[6] as it can be used to explore not only greater opportunities for reducing emissions[7] but also the achievement of synergies with other development goals.[8] However, there are ethical and political cautions in seeking transformation which are further discussed in Chap. 10.

Transition and transformation are known as backcasting scenarios, as they pick a desired end-result and seek to strategise how it could be achieved.[9] This of course has relevance in the case of transition as it involves the goal of a steep decline in GHG emissions globally over the course of the twenty-first century. As noted by Robinson (2003: 842), this is particularly important in the case of sustainability, where the most likely futures may not be the most desirable. Backcasting scenarios involve defining the desired end goal and the different pathways for its achievement, including the policy measures required. It is frequently implemented as a democratic participatory process with stakeholders affected by the outcomes.

A Review of Transition and Transformation Studies

Transition and transformation studies have been produced by research organisations, policy think-tanks, states, cities and NGOs. Here we seek to review some of the key outcomes, by reviewing the overall conclusions with illustration from some of the prominent studies that have been produced. The non-governmental report *Who's Getting Ready for Zero?* produced by the Centre for Alternative Technology and Track 0 (CAT/Track 0 2015) maps out how different actors at national, regional and city levels are already modelling the elimination of GHGs on timeframes compatible with 2°C. The report reviews over 100 scenarios showing that decarbonisation scenarios have been created for a wide range of countries, including sixteen of the world's largest emitters, emitting nearly 75% of the world's carbon emissions. The report noted that although a proportion of the Intended Nationally Determined Contributions (INDCs) submitted to date under the UNFCCC process do contain a long-term objective or perspective, there remain many countries that have not yet prepared scenarios or strategies aligning short-term mitigation and development goals with longer-term 2050 timeframes. The report highlights a number of common observations in 'deep decarbonisation', 'low carbon' and 'zero carbon' scenario studies: (1) it is necessary to consider lifestyles, (2) the required technologies are available, (3) strategies must be multisectoral, (4) there are many co-benefits available, and (5) climate action is pro-human development and pro-fairness.

A major World Bank report entitled *Decarbonizing Development: Three Steps to a Zero Carbon Future* (World Bank 2015) focuses on common economic concerns and prescriptions including cost effectiveness, carbon prices, technology, behaviour, finance, sectoral approaches and achieving co-benefits. These largely techno-centrist approaches to mitigation are recommended as 'action on the four pillars of decarbonisation', that is, decarbonised electricity, electrification, efficiency and preservation of carbon sinks. *The New Climate Economy Report* of The Global Commission on the Economy and Climate (2014), noted that the IPCC's Fifth Assessment Report has reviewed the global economic literature on the costs of climate action, concluding that costs are likely to be small, at around 1.7% (median) of baseline global GDP in 2030. The report describes this as 'background noise' when compared with the strong growth that the global economy is likely to experience. The report found that model-based assessments tend not to include many of the benefits, and that costs tend to disappear when existing economic inefficiencies and the benefits of action are taken into account. In supporting these conclusions, the 'Wind-Water-Sun' study of Delucchi and Jacobson (2011) determined that it is both technically and economically feasible to convert the entire global energy system to 100% renewables by 2050. As a high-profile study that generated media interest, it is detailed in Box 5.4.

Box 5.4: The 100% Renewable Wind-Water-Sun Global Energy System in 2050

The Delucchi and Jacobson Wind-Water-Sun study (Delucchi and Jacobson 2011) became a high-profile and controversial technology feasibility study of the potential to convert the entire global energy system to renewables, the 'WWS energy system'. Published in the journal *Energy Policy* in 2011, the study by a transportation expert and an atmospheric scientist from UC Davis and from Stanford University found that it is both technically and economically feasible to convert the entire global energy system to 100% renewables by 2050 for all energy consuming purposes: electric power, transportation and heating and cooling. This would involve a series of technical measures from interconnection and electricity storage to hydrogen production for the supply of different forms of energy, and balancing variability in production from renewables. They concluded that the cost of energy in a 100% WWS would be similar to the cost today, meaning that the barriers to conversion to a WWS powered world are primarily social and political and not technological or economic. Such claims have been debated since, with Williams (2013) suggesting that there have been limitations in such studies to date including the depth of the coverage of energy consuming sectors, physical and resource constraints and geographic scale. Nevertheless, in refining the scale to California, Williams found that such claims were indeed supported. When energy efficiency and decarbonisation of power generation are supplemented by increased electrification, the deep decarbonisation moves California from an oil economy to an electric economy.

The 'Post Carbon Pathways' project of the Centre for Policy Development and the Melbourne Sustainable Society Institute (Wiseman et al. 2013) aimed to learn from the thinking on what they envisaged as the most promising decarbonisation strategies and ways of overcoming barriers to implementation. It was informed by a review of 18 of the most prominent governmental and non-governmental transition strategies such as the German Advisory Council on Global Change's *World in Transition* (2011), Al Gore's *Our Choice* (2009), *Zero Carbon Britain 2030* (2010), the European Commission's *Low Carbon Roadmap* (EC 2011) and the Government of China's *12th Five-Year Plan & Climate Change White Paper* (2011). They came to the conclusion that public awareness and vision are necessary, and that key roadblocks include climate science denial, vested interests, unsustainable consumption, lock-in and finance.

One of the most prominent decarbonisation plans mentioned above, the EU *Roadmap for moving to a competitive low-carbon economy in 2050* was released by the European Commission in 2011 (EC 2011). Moving beyond the common short-term framework to 2020 at the time, it set out the plan to meet the long-term target of reducing domestic EU GHG emissions by 80%–95% by mid-century. This target was in line with the

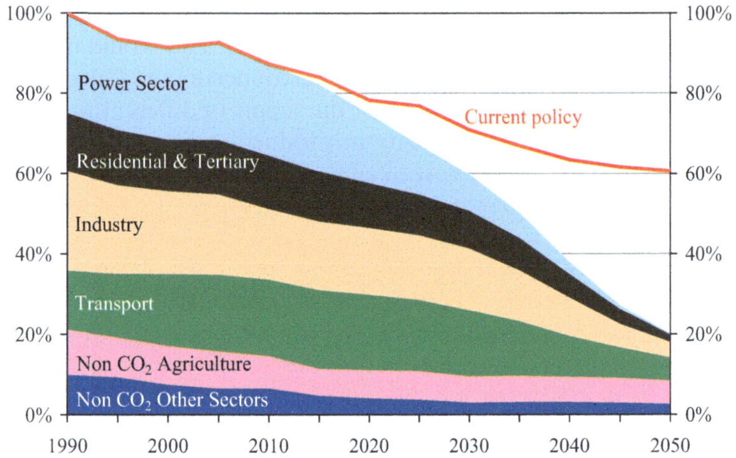

Fig. 5.1 EU GHG emissions towards an 80% domestic reduction (100% = 1990). Source: EU low-carbon roadmap (EC 2011: 5) http://eur-lex.europa.eu, © European Union, 1998–2017

conclusions of the IPCC on the emissions reductions required to keep temperature increase below +2°C, and agreed by world leaders at the Copenhagen and Cancun CoPs.[10] It addressed the economic sectors of power generation, residential and commercial services (tertiary), industry, transport and agriculture. Figure 5.1 illustrates the reductions envisaged in each of these sectors in the EU to 2050.

As a technical feasibility study, it sought to provide a long-term framework to assist EU member states in policy development. It showed that domestic emission reductions of the order of 40% and 60% below 1990 levels would be 'cost-effective' by 2030 and 2040. It concentrated on the use of energy efficiency and renewable energy policy, and early deployment of technologies, such as various forms of low-carbon energy sources, carbon capture and storage, smart grids and hybrid and electric vehicle technology, to ensure large-scale penetration later on. It described synergies with other sustainability objectives such as reduction of oil dependence, enhancing competitiveness of Europe's automotive industry as well as health benefits, especially improved air quality in cities. While establishing the technological path to a decarbonised Europe is a valuable and necessary goal, it is a narrow focus in transition. Because it is framed as a 'low-carbon economy', it is essentially silent on sustainable development pathways that change our overall approach to development. It does not address the social, cultural and political aspects of development and technological transition, and the political economy of these changes in the future. From hybrid and electric cars, passive housing and renewable energy, to behavioural changes, it is radical technologically. It is also commendable that it refers to changes towards more healthy diets to reduce emissions, with more vegetables and less meat, but in essence it is more of a technological transition. Nevertheless, as the EU is a significant player in climate change negotiations, it has helped shift global political opinion towards the feasibility of the low-carbon future, and spurred EU member states to engage substantively with the long-term change required. Further details on plans in individual countries, including the EU, can be found in Chaps. 6 and 7.

The 'Tellus Institute scenarios' (Raskin et al. 2010) offer a useful contrast to the technological focus of most decarbonisation scenarios and planning. The scenarios were a prominent study continuing a programme spanning decades of exploring alternative global futures. Conducted by the Tellus Institute, the Global Scenario Group and the Stockholm Environment Institute, it was pioneering scenario work that could be described as exploring both incremental and also transformative change for a sustainable, and not just a low-carbon future.[11] Exploring possible pathways to sustainability over the twenty-first century, they include thematic impacts and outcomes such as energy requirement, carbon emissions, toxic chemical load, food and hunger, water stress and human wellbeing. The scenarios highlight the risks of conventional development approaches. The 'Market Forces' and 'Policy Reform' scenarios emerge gradually from the dominant forces governing world development at the time of the study. The second two scenarios evolve from a fundamental restructuring of the global order: fragmentation in 'Fortress World' and positive transformation in 'Great Transition'. In 'Fortress World' an authoritarian path is the reactive response to multiple security, environmental and social crises, with a decline in the fortunes of humanity occurring, a prescient portent of current dynamics. In the 'Great Transition' scenario a fundamental and urgent shift in the development paradigm manifests towards enhancing human wellbeing and preserving the ecosphere achieved through both strategic and values change.

The 'Great Transition' and 'Policy Reform' scenarios both show a drop in global carbon dioxide emissions to zero before 2100. The study used a 'Quality of Development Index' (QDI) to measure change in three

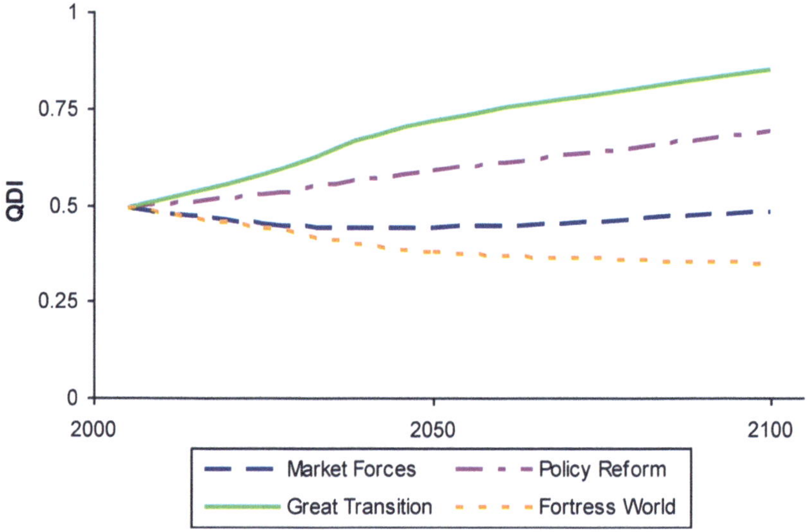

Fig. 5.2 Global quality of development index from the Tellus scenarios. Source: The Tellus Institute scenarios in Raskin et al. (2010: 2633)

key aspects of sustainable development: human wellbeing, community cohesion and environmental protection. In Fig. 5.2 we can see the superior performance of both the 'Great Transition' and 'Policy Reform' scenarios in sustainability outcomes. The holistic sustainability approach taken by the Tellus scenarios is useful in moving beyond the technological transition. It also has clear implications for political economy, as the style and substance of governance dictate which type of outcomes arise.

A Critical Analysis of Looking into the Future, from Transition to Transformation

From the sections above on techniques to look into the future, on environmental scenarios and on emissions scenarios, we got impressions of the approaches that are used, what aspects are important to look at and what the implications are. They offer a toolkit to look at future change and to appraise transition. For long-term environmental and emissions challenges, scenario analysis is the chosen approach to understand the different possible outcomes. Many scenario studies have relied solely on the technological and economic variables, those variables that are amenable to energy and economy models. Some studies try to integrate narrative story-lines of how social, cultural and political drivers change with quantitative modelling of the economy and technology. As the significance of these drivers is acknowledged, the place of the social sciences and political economy is raised as a critical lens. This requires more than energy technology and the economics of change, it requires us to understand sustainable development pathways and what they mean for society. While sustainable development can sometimes be criticised as 'greenwashing' if it is used to legitimise developments that are not in the direction of sustainability (Sathaye et al. 2007: 697), it does provide a useful perspective on what type of change is occurring. It must be politically defined if we are to use this to interrogate the pathways on which we are developing, who is benefitting and what it means for the environment and for future generations? It calls for an active engagement in scenario studies of those with expertise in political economy, and more complete coverage of social, cultural and political driving forces in scenario studies. How this relates to the environment we wish to preserve, the kind of society we want and how we will get there, are questions that are now of pressing need for the social sciences and political economy to consider.

The 'low-carbon transition' could be seen as a popular catchphrase for responses to climate change, for 'deep decarbonisation'[12] and avoidance of dangerous levels of warming of the climate. Transition has thus been a byword for technological change that decarbonises the energy system in the long term, with transition studies at national, regional and city levels aiming to elucidate what this will mean. If we wish to continue with the current concept of civilisation, with a continued function for energy-consuming technology, then a technological transition is necessary to ensure that the required energy services are delivered without adding to the stock of GHGs in the atmosphere. Low-carbon transition studies are converging towards a common set of conclusions on what this technological challenge means:

1. the low-carbon transition is technically and economically feasible,
2. transition comes with multiple co-benefits,
3. replacement of fossil energy systems with renewables, increased electrification of energy consumption and strong pursuit of energy efficiency, are identified as the necessary elements of technological change.

A critical review of low-carbon transition and energy scenarios was conducted by Soderholm et al. (2011: 1113). They note 'a particularly salient weakness of previous low-carbon quantitative energy scenarios is that they tend to adopt a rudimentary approach when it comes to issues about politics, institutions and governance'. Many backcasting studies represent what Hughes et al. (2009) term 'technical feasibility studies' rather than comparative policy analyses. However, more significantly than this point, what needs to be highlighted is the crucial gap in knowledge as transition studies do not adequately account for sustainable development paths and the wider influence of social, cultural and governance factors on how these sustainable development paths unfold. Such transition studies may then be putting the 'cart before the horse'. It is known that the barriers to technological transition are predominantly social and political (Delucchi and Jacobson 2011: 1154). More importantly, it is also known that transition must begin with sustainable development pathways (Sathaye et al. 2007), which are also predominantly social and political challenges. There is a need to move from the techno-centrism and techno-optimism of technical feasibility studies, to transition and transformation of society and sustainable development pathways. Social justice, empowerment and equality, environmental protection, ethics and enhanced human wellbeing are central to this task. The political economy of social and political arrangements could be seen as the next step in the challenge of understanding and implementing a post-carbon society. The EU low-carbon roadmap (2011) involved an online public consultation and this is a step in the right direction in terms of democratic participation.

In order to fully understand the technological transition it is also useful to acknowledge a debate on potential excessive techno-optimism that has emerged. Kevin Anderson and Alice Bows of the Tyndall Centre for Climate Change in the UK are highly sceptical of mainstream conclusions on the cost and feasibility of the technological transition. They argue that how climate change science is conducted, communicated and translated into policy must be radically transformed if 'dangerous' climate change is to be averted. Anderson and Bows state that: 'behind the cosy rhetoric of naively optimistic science and policy, there is little to suggest that existing mitigation proposals will deliver anything but rising emissions over the coming decade or two' (Anderson and Bows 2012: 639). Anderson followed this line of argument in a commentary in *Nature Geoscience* (Anderson 2015: 1) stating that the conclusions on the 'economically feasible' transition away from fossil fuels to keep warming to less than

2°C involve: 'unrealistically early peaks in global emissions, or through the large-scale rollout of speculative technologies intended to remove CO_2 from the atmosphere yielding so-called negative emissions'. Regardless of how politically palatable Anderson's view is, if the mainstream consensus is indeed excessively optimistic then more rapid emissions reductions, and fundamental transformations of how wealthy countries consume energy, natural resources and consumer products, then become urgently necessary.

The non-governmental report *Who's Getting Ready for Zero?* (CAT/Track 0 2015), is useful in summarising the breadth of the mainstream literature related to decarbonisation and transition, but given the existing conclusions throughout the literature, it is what is missing rather than what is included that is more instructive. Considering the outcomes of the report it is clear that the focus remains techno-economic, on productive economic sectors, technologies, co-benefits and individuals. It focuses on aspects of the technological transition further reflecting the blindspots in the transition literature in general and transition policy worldwide. This can be summarised as sustainable development pathways, development models, social, cultural and governance drivers and policy synergies, the messy real world of policy that transition must address, not in idealised or optimised form. Such a report can function as a political catalyst to action, and assist modellers in synthesising conclusions, but for a fully integrated conception of the future a more holistic transdisciplinary development approach is now necessary. This involves a move from transition scenarios to transformation scenarios. The Tellus Institute scenarios of Raskin et al. (2010) are a useful contrast here, as they show not only a low-carbon transition but elements of a sustainable society. They highlight that it is the type of policy that is implemented, and more significantly, the cultural and political changes in the 'Great Transition' and 'Policy Reform' scenarios, that lead to these more desirable outcomes.

The 'transition management' literature based in the field of co-evolutionary economics and in innovation theory, has developed a way of understanding socio-technical system change and how the process of 'managing' system innovation and transformation could be achieved. Patterson et al. (2016) point to a variety of other conceptual approaches used to understand and analyse societal transition or transformation processes. In addition to the socio-technical transitions approach this includes 'social-ecological systems', 'sustainability pathways', and 'transformative adaptation'. All of these approaches offer potential assistance in furthering our understanding of transition and transformation. Nevertheless, from a broad analytical and strategic policy perspective, it is necessary to transcend disciplinary boundaries and include multiple approaches from different perspectives. This could lead to an integrated and holistic conception of the future, and not just limit inquiry to techno-economics. It could be argued that the lack of sufficient inclusion of the social sciences in the analytical methods applied is the reason for the absence of a focus on these more fundamental measures of mitigation. They are discussed only very rarely, and are usually not substantively included in the analytical focus, or in the development and implementation of policy for mitigation and transition. There is a continuing reliance on techno-economic measures that have failed to drive change at the speed or scale required. These involve profound issues of politics and power. The many inter-related development decisions we are taking today from economic to social policy, and from transition to energy and environmental protection, have long-term implications for how society is organised, for whose voices are heard and for whom there are benefits and who loses out. The critique of political economy can be a useful lens on the state-market-society relationships, and the power relations within them, as we move towards a post-carbon and sustainable world.

Conclusion

This chapter has discussed a crucial area in transition and transformation towards post-carbon and sustainable societies, the use of scenarios. The history of scenario analysis and planning spans decades, and they have been used for a range of purposes, from academic research to policymaking and conflict resolution. Used prominently in energy, environment and greenhouse gas emission studies, many only address economic and technological change, while some have sought to integrate social, cultural and political drivers. In transition and transformation, a related problem sees the dominance of technical feasibility studies addressing the technological approach to low-carbon transition. There are examples of more integrated approaches such as the Tellus Institute scenarios that address both sustainability and low-carbon transformation. The increase in low-carbon and zero-carbon transition studies is converging towards a common set of conclusions: the low-carbon transition is technically and economically feasible, transition comes with multiple co-benefits, electrification, renewables and efficiency are key and the barriers are social and political. While the technological transition to a low-carbon future may be technically and economically feasible according to the mainstream view, there are also those who question the validity of this conclusion as assumptions may be excessively optimistic (Anderson and Bows 2012; Anderson 2015). If this view is more valid, it will require even more urgent action, and a fundamental re-thinking of how energy and resources are consumed among the more wealthy. There are also questions about the development and sustainability consequences of transition in general. From the political economy discussion in Chap. 4, we know that a development discussion is about politics and power, who participates in this development, in what ways and for the benefit of whom. This is what Robinson referred to as 'profound issues of opportunity, distribution, material needs, consumption and empowerment' in moving towards a sustainable society, which in turn are inseparable from issues of social and political organisation and governance (Robinson 2004: 379). The studies produced, and the changes underway, have profound social and political implications. From both a research and policy perspective, a greater inclusion of the social sciences would assist in understanding these changes. Crucially, political economy could give valuable perspectives on the type of state-market-society relationships that emerge.

Notes

1. A 'model' here is defined differently to that in previous chapters pertaining to political economy models or frameworks for development. Models here are deterministic quantitative representations of how the techno-economic system functions that allow simulation under different conditions. Soderholm et al. (2011: 1106) described them as follows: 'Traditionally quantitative assessments have rested on the use of two types of models that address the interactions between the energy sector, the economy and the environment: top-down and bottom-up models. They differ mainly with respect to the emphasis placed on detailed, technologically based treatment of the energy system (bottom-up), and theoretically consistent descriptions of the economy (top-down).'
2. They can be arbitrary in the sense that the quantitative variables modelled may not be the most significant in driving the system. The approach often uses past outcomes that are statistically analysed and used to forecast or project future outcomes. But as the models deal with complex non-linear systems, there is no guarantee that past relationships hold. This was evidenced by the economic and financial crisis from 2008 which confounded forecasters in many countries and their projections of the economy, energy and emissions.

3. From human activities such as fossil fuel combustion for power, transport and heat, livestock methane emissions, land use change and deforestation and industrial process emissions.
4. Avoiding 'Dangerous Anthropogenic Interference' with the climate system or 'DAI' is the stated objective of the UN Framework Convention on Climate Change (UNFCCC) as detailed in Article 2.
5. The mixture of state, market and civil society.
6. It also needs to be recognised that rapid and dramatic transformation can be traumatic, so it may not be ethical or desirable in all its hues. However, adopting a status quo approach is likely to lead to severely negative outcomes.
7. Transformation not only involves facilitating the techno-economic transition, but change in the crucial social, cultural and governance factors and in the underlying development path. Changing the underlying development path, known as 'immaterialisation', is deeper than technical efficiency and decarbonisation, as it can prevent the consumption of energy and materials at source rather than 'end-of-pipe' technological solutions (O'Mahony and Dufour 2015: 418). This increases the breadth of the menu of options to reduce emissions beyond the techno-economic.
8. Synergies with other development goals can be achieved where the development path reduces emissions in parallel to enhanced human development, improved human wellbeing, improved environmental quality and/or increased and more sustainable economic growth.
9. Rather than beginning from the present and then forecasting predicted outcomes (forecasts), projecting a continuation of the past (projections) or exploring possible alternative future worlds (exploratory scenarios).
10. Meetings of the Conference of the Parties to the UNFCCC held in Copenhagen in 2009 and in Cancún in 2010.
11. A low-carbon future might reduce global emissions sufficiently below the level required to limit temperature increase to +2°C or even +1.5°C of global warming, and avoid 'dangerous' climate change, but it would not necessarily be 'sustainable' or lead to a sustainable society. A low-carbon future could still entail significant environmental damage if the environment is not adequately protected, or indeed social collapse if human wellbeing, social justice, inequality and poverty are not addressed. A sustainable society would not just be 'low-carbon', it would need to address the other social, environmental and economic issues at the same time.
12. 'Low carbon', 'zero carbon', 'post carbon' and 'deep decarbonisation' are often used synonymously. We have preferred post carbon as it suggests a move to a world that has transformed completely beyond the limitations of the fossil fuel based society and economy.

References

Alcamo, J., ed. 2008. *Environmental Futures. The Practice of Environmental Scenario Analysis. Developments in Integrated Environmental Assessment 2*. Amsterdam: Elsevier.

Anderson, K. 2015. Duality in Climate Science. *Nature Geoscience* 8: 898–900.

Anderson, K., and A. Bows. 2012. A New Paradigm for Climate Change. *Nature Climate Change* 2: 639–640.

Carpenter, S.R., P.L. Pingali, E.M. Bennett, and M.B. Zurek. 2005. *Ecosystems and Human Well-being: Scenarios, Volume 2. Millennium Ecosystem Assessment*. Washington, DC: Island Press.

CAT/Track 0. 2015. *Who's Getting Ready for Zero?* Centre for Alternative Technology and Track 0. Accessed 24 March 2017. http://zerocarbonbritain.org/images/pdfs/wgrz-full-report.pdf

Chappin, E.J.L., and Andreas Ligtvoet. 2014. Transition and Transformation: A Bibliometric Analysis of Two Scientific Networks Researching Socio-technical Change. *Renewable and Sustainable Energy Reviews* 30: 715–723.

De Jouvenel, H. 2000. A Brief Methodological Guide to Scenario Building. *Technological Forecasting and Social Change* 65: 37–48.

Delucchi, M.A., and M.Z. Jacobson. 2011. Providing All Global Energy with Wind, Water, and Solar Power, Part II: Reliability, System and Transmission Costs, and Policies. *Energy Policy* 39: 1170–1190.

EC. 2011. A Roadmap for Moving to a Competitive Low Carbon Economy in 2050. Communication from the European Commission COM(2011) 112. eur-lex.europa.eu/legal-content/EN/TXT/PDF/?uri=CELEX:52011DC0112&from=EN

Fisher, B.S., N. Nakicenovic, K. Alfsen, J. Corfee Morlot, F. de la Chesnaye, J.-Ch. Hourcade, K. Jiang, et al. 2007. Issues Related to Mitigation in the Long Term Context. In *Climate Change 2007: Mitigation. Contribution of Working Group III to the Fourth Assessment Report of the Intergovernmental Panel on Climate Change*, ed. B. Metz, O.R. Davidson, P.R. Bosch, R. Dave, and L.A. Meyer. Cambridge: Cambridge University Press.

Fleurbaey, M., S. Kartha, S. Bolwig, Y.L. Chee, Y. Chen, E. Corbera, F. Lecocq, et al. 2014. Sustainable Development and Equity. In *Climate Change 2014: Mitigation of Climate Change. Contribution of Working Group III to the Fifth Assessment Report of the Intergovernmental Panel on Climate Change*, ed. O. Edenhofer, R. Pichs-Madruga, Y. Sokona, E. Farahani, S. Kadner, K. Seyboth, A. Adler, I. Baum, S. Brunner, P. Eickemeier, B. Kriemann, J. Savolainen, S. Schlömer, C. von Stechow, T. Zwickel, and J.C. Minx. Cambridge, UK and New York, NY: Cambridge University Press.

Hughes, N., J. Mers, and N. Strachan. 2009. Review and Analysis of UK and International Low Carbon Energy Scenarios. A Joint Working Paper of the

UKERC and the EON.UK/EPSRC Transition Pathways Project. http://www.ukerc.ac.uk/asset/CF4C4FA6%2D3977%2D4138%2D9CE618FA1D842534/

IPCC. 2012. *Managing the Risks of Extreme Events and Disasters to Advance Climate Change Adaptation*. Special Report of the Intergovernmental Panel on Climate Change. Cambridge: Cambridge University Press.

Kaya, Y. 1990. Impact of Carbon Dioxide Emission Control on GNP Growth: Interpretation of Proposed Scenarios. Paper presented to the *IPCC Energy and Industry Subgroup, Response Strategies Working Group*, Paris (mimeo).

Meadows, D.H., D.L. Meadows, J. Randers, and W.W. Behrens III. 1972. *The Limits to Growth: A Report for the Club of Rome's Project on the Predicament of Mankind*. New York: Universe Books.

Midttun, A., and T. Baumgartner. 1986. Negotiating Energy Futures: The Politics of Energy Forecasting. *Energy Policy* 14: 219–241.

Nakicenovic, N., J. Alcamo, G. Davis, B. de Vries, J. Fenham, S. Gaffin, K. Gregory, et al. 2000. *Special Report on Emissions Scenarios. Working Group III, Intergovernmental Panel on Climate Change (IPCC)*. Cambridge: Cambridge University Press.

Nielsen, S.K., and K. Karlsson. 2007. Energy Scenarios: A Review of Methods, Uses and Suggestions for Improvement. *International Journal of Global Energy* 27 (3): 302–322.

O'Mahony, T. 2013. *The Future of Ireland to 2020. Putting the Story in Emission Scenarios*. Saarbrücken: Lambert Academic Publishing.

———. 2014. Integrated Scenarios for Energy: A Methodology for the Short Term. *Futures* 55: 41–57.

O'Mahony, T., and J. Dufour. 2015. Tracking Development Paths: Monitoring Driving Forces and the Impact of Carbon-free Energy Sources in Spain. *Environmental Science and Policy* 50: 62–73.

O'Mahony, T., P. Zhou, and J. Sweeney. 2013. Integrated Scenarios of Energy-related CO_2 Emissions in Ireland: A Multi-sectoral Analysis to 2020. *Ecological Economics* 93: 385–397.

O'Neill, B., E. Kriegler, and K. Ebi. 2015. The Roads Ahead: Narratives for Shared Socioeconomic Pathways Describing World Futures in the 21st Century. *Global Environmental Change* 42: 169–180.

Patterson, J., K. Schulz, J. Vervoort, S. van der Held, O. Widerberg, C. Adlere, M. Hurlbert, K. Anderton, M. Sethih, and A. Barauja. 2016. Exploring the Governance and Politics of Transformations Towards Sustainability. *Environmental Innovation and Societal Transitions*. http://www.sciencedirect.com/science/article/pii/S2210422416300843

Raskin, P., C. Electris, and R.A. Rosen. 2010. The Century Ahead: Searching for Sustainability. *Sustainability* 2: 2626–2651.

Raskin, P., F. Monks, T. Ribeiro, D. van Vuuren, M. Zurek, A.A. Concheiro, and C. Field. 2005. Global Scenarios in Historical Perspective. In *Ecosystems and Human Well-Being: Scenarios, Findings of the Scenarios Working Group Millennium Ecosystem Assessment Series*, ed. Steve Carpenter, Prabhu Pingali, Elena Bennett, and Monika Zurek, 35–44. Washington, DC: Island Press.

Robinson, J. 2003. Future Subjunctive: Backcasting as Social Learning. *Futures* 35: 839–856.

———. 2004. Squaring the Circle? Some Thoughts on the Idea of Sustainable Development. *Ecological Economics* 48: 369–384.

Sathaye, J., A. Najam, C. Cocklin, T. Heller, F. Lecocq, J. Llanes-Regueiro, J. Pan, et al. 2007. Sustainable Development and Mitigation. In *Climate Change 2007: Mitigation. Contribution of Working Group III to the Fourth Assessment Report of the Intergovernmental Panel on Climate Change*, ed. B. Metz, O.R. Davidson, P.R. Bosch, R. Dave, and L.A. Meyer. Cambridge: Cambridge University Press.

Soderholm, P., R. Hildingsson, B. Johansson, J. Khan, and F. Wilhelmsson. 2011. Governing the Transition to Low-carbon Futures: A Critical Survey of Energy Scenarios for 2050. *Futures* 43: 1105–1116.

Turner, G.H. 2008. A Comparison of the Limits to Growth with 30 Years of Reality. *Global Environmental Change* 18: 397–411.

UNEP. 2012. *Global Environmental Outlook 5, Environment for the Future We Want.* Summary for Policymakers. United Nations Environment Programme, downloaded from http://web.unep.org/geo/assessments/specialized/geo-5-summary-policy-makers

———. 2016. Rate of Environmental Damage Increasing Across the Planet but There Is Still Time to Reverse Worst Impacts if Governments Act Now, UNEP Assessment Says: Landmark UNEP Assessment Puts State of the World's Environment under the Microscope. United Nations Environment Programme News Centre. http://www.unep.org/newscentre/default.aspx?DocumentID=27074&ArticleID=36180

Williams, J.H. 2013. The Technology Path to Deep Greenhouse Gas Emissions Cuts by 2050: The Pivotal Role of Electricity. *Science* 335 (6064): 53–59.

Wiseman, J., T. Edwards, and K. Luckins. 2013. Post Carbon Pathways, Towards a Just and Resilient Post Carbon Future. Learning from Leading International Post-carbon Economy Researchers and Policy Makers. Melbourne Sustainable Society Institute, University of Melbourne CPD Discussion Paper.

World Bank. 2015. *Decarbonizing Development: Three Steps to a Zero-Carbon Future*. Washington, DC: World Bank.

PART III

Pathways in Developed and Developing Countries

CHAPTER 6

Development and Sustainability in the Wealthiest Regions: Taking the High Road?

Introduction

Since the Rio Earth summit in 1992 and the establishment of the UN Framework Convention on Climate Change (UNFCCC), which entered into force in 1994 following ratification by 50 countries, an international process has been in place focusing the attention of policy makers at the national level on the need to 'mitigate' or reduce GHG emissions. This initial short-term perspective later evolved into considering longer-term low-carbon development pathways as the emissions trajectories required to avoid 'dangerous climate change' throughout the twenty-first century became clearer. Deciding what were 'dangerous' levels of interference with the climate was a political process, informed by science, that crystallised around +2°C change on the pre-industrial average global temperature, and later also to pursuing efforts to limit the increase to +1.5°C as committed to in the Paris Agreement. The principle of *common but differentiated responsibilities* (CBDR) between developed and developing countries enshrined in the UNFCCC from the beginning allowed the latter to resist binding emissions targets. It acknowledged they had less historical responsibility for the emissions that led to climate change, and also their need to develop economically. The first international agreement to set binding targets, the Kyoto Protocol which entered into force in 2005, applied only to developed countries (the so-called Annex I countries).[1] This divide has so marked the UNFCCC process as it has evolved that

P. Kirby, T. O'Mahony, *The Political Economy of the Low-Carbon Transition*, International Political Economy Series,
DOI 10.1007/978-3-319-62554-6_6

Torney states that 'the defining feature of contemporary global climate politics concerns the divide between industrialized and developing countries' (Torney 2015: 79).

Since Kyoto, many developments have moved the narrative forward, galvanising public, political and business opinion worldwide. Fossil-fuelled 'brown-growth' became less desirable with each passing year. The economic feasibility of transition and 'greener growth', and the many benefits of such action, became clearly evident. Meanwhile, the science of climate change became even more unequivocal, and temperature increase and climate impacts became more severe, increasingly affecting developing countries. Carbon-intensive economic growth propelled major emerging economies such as China, India, Brazil and Mexico to join the US, the EU, Russia and Japan among the top ten global emitters of GHGs through development that increased the ecological debt. Acceptance of the limited global carbon budget available into the future, crystallised from advances in the science. It became evident that if the developing world copied the emissions-intensive development paths of the industrialised world, a stable climate and development gains would be lost to history. It then became an acceptable narrative that developing countries would also now adopt emissions reduction targets, compatible with sustainable development and climate protection. This was helped by the commitment of developed countries at the Copenhagen climate summit in 2009 to contribute up to $30bn to developing countries between 2010 and 2012, and $100bn annually from 2020 onwards, to help them enhance action to reduce emissions. Thus, at the Durban climate summit in 2011, developing countries agreed that the successor to the Kyoto Protocol would, in some form, be binding for all countries, not just developed ones.

This steady if slow process of global climate diplomacy has, therefore, helped place the issue of curbing global warming and dealing with the impacts of climate change, mitigation and adaptation respectively, high on the political agenda of countries worldwide. With different amounts of political commitment, countries are developing their own pathways to a more sustainable future, which is increasingly being defined as becoming a low-carbon economy and sustainable society by 2050. The purpose of Section III is to examine what they are doing and what lessons can be learned. Which are the leaders and which the laggards in dealing with this historically daunting challenge? This chapter looks at two groups of countries, all of which have prioritised conventional economic development: the 'developed world' countries and regions of the EU, the US and Japan and the 'emerging economies' of China, India, Brazil and Mexico. The emerging economies have shown some of the highest economic growth rates in the world up to recently, and have also begun showing leadership on climate change in different ways. As its title indicates, the chapter looks at how, and indeed if, countries are seeking to 'take the high road' by combining development and sustainability in the task of reducing emissions. As the chapter makes clear, different countries have taken very different approaches. The next chapter will focus on countries throughout the developing world, illustrating the different ways in which they are seeking to combine development with transition and sustainability. It is clear that this 'wicked problem' presents itself very differently in different countries: from their resources, economy and levels of development, to their social, cultural and political characteristics and from their environmental legacy to their technological potential. Finally, each chapter will seek to identify the political economy model that is structuring the response to climate change, the mix of state, market and society involved.

The 'Developed World': The EU, the US and Japan

This group has for long been the source of the highest emissions per capita (Table 6.1), and therefore been faced with the initial challenges of how to reduce emissions, but each has approached the problem very differently. Table 6.1 offers some selected social, environmental and economic indicators of how well they are succeeding in combining development with sustainability for the developed alongside the 'emerging market' countries. The Ecological Footprint (EF) is increasingly used as a measure of society's pressures on the global ecosystem.[2]

The EU, Diverse Countries with a Diversity of Opportunities

The European Union, a bloc of 28 European countries (to be reduced to 27 if and when the UK leaves), evolved from a trade bloc into the modern politico-economic union, with both centralised and de-centralised decision-making. The EU28 spans a population of 510 million, from the Nordics to the Baltics and from the 'industrialised' west to the Mediterranean, and also a range of economic, social, cultural, political and environmental approaches. What unites them is cooperation for mutual benefit, including common positions on core issues. 'Sustainable development', including action on climate change, has been successively strengthened in EU treaties so that the Union is regarded as a pioneer of environmental policy,

Table 6.1 Selected development indicators

Country/Region	*GHG tonnes per capita 2014*[a]	*GDP per capita 2014 $*[b]	*Income inequality by Gini coefficient 2014*[c]	*Life satisfaction 2013*[d]	*EF per capita global hectares 2012*[e]	*How many planets?*[f]
EU28	8.72	37,553	30.9	6.1	4.5	2.8
USA	21.55	54,599	39.4	7.2	8.2	4.8
Japan	10.74	39,387	33.0	6.1	5.0	2.9
China	8.13	13,440	42.2	5.3	3.4	2.0
India	2.44	5,678	35.2	4.3	1.2	0.7
Brazil	5.10	16,192	52.7	6.5	3.1	1.8
Mexico	5.99	17,363	48.1	6.2	2.9	1.7

[a]OECDstat and EUROSTAT excluding LULUCF
[b]World Bank database, in Purchasing Power Parities at current prices
[c]OECDstat and EUROSTAT, Japan 2012 data
[d]World database of happiness, EU27 data
[e]footprintnetwork.org, EU 2010 data by EEA
[f]footprintnetwork.org

including efforts on water and air pollution, nature conservation and environmental protection. In climate change policy and sustainable energy it has excelled, and the EU is perceived internationally as a leader. It has been a champion of the UNFCCC process since its inception. Among developed nations, it effectively carried the Kyoto Protocol when other key players such as the US, Canada and Australia chose to forego their responsibilities, and also the inherent opportunities. The EU exceeded its Kyoto target of a −8% reduction in GHG emissions by 2008 to 2012 (on 1990 levels), and is continuing to drive emissions down towards 2030 targets of −40% and 2050 of −80% to −95%.

Key policies include the EU *Climate and Energy Framework* to 2030 and binding national targets and directives on emissions reduction, renewable energy, energy efficiency and the EU Emissions Trading Scheme (ETS) in the case of the industry and power sectors. The EU *low carbon roadmap* to 2050 is a key plank in transition towards a low-carbon future, to establish a 'cost-efficient pathway' to reach the −80% target, with milestones of −40% below 1990 by 2030 (as the target already endorsed in the 2030 framework) and −60% below by 2040. It has also begun to note wider development issues such as diet, suggesting the need for a reduction in meat consumption; it is however, predominantly a technological transition. The EU has been a global leader in renewable energy research and implementation, but recent austerity policies are now affecting investment and support. The leadership role is consequently being lost to China. The EU envisages a 'cost-efficient pathway' but this can be a dubious concept when we consider the full environmental and social costs of fossil fuel consumption.[3] Examples of problematic sectors that run contrary to emissions objectives include coal-fired power generation in Poland and Greece, protections for expansion of dairy and livestock agriculture in Ireland and France, and growth in transport emissions and the aviation sector across the EU.

The EU emissions pie is carved up between the 'big emitters' that come under the EU ETS and the 'non-ETS sectors' including all other GHG sources such as those from commercial services, public services, transport, residential, agriculture and land use and forestry, all of which are national responsibilities. While this 'economic' classification of 'sectors' is useful, an *integrated development approach* that goes beyond 'economic sectors', and improved mainstreaming of climate policy, is needed for sustainable development. Promising approaches include efforts towards resource efficiency, green growth and the circular economy, but the successful achievement of Kyoto may distract from increasing material consumption levels and their associated emissions, which are not compatible with a sustainable development path. The territorial based accounting of emissions for the Kyoto Protocol shows a drop in emissions, but consumption-based accounting, which includes emissions embodied in products imported through trade, shows that material consumption and related emissions have not actually dropped (EEA 2015). As industrialised economies move to services, emissions are effectively outsourced to developing countries, which provide most manufactured goods. As material consumption has not decreased, emissions are embodied in traded goods but do not show up in EU emissions data. Therefore a large portion of emissions growth in the production sectors of the emerging economies is attributable to the consumption of the affluent outside their jurisdiction.

This could be described as an accounting trick, and it needs to be addressed for genuine low-carbon development. It becomes an even more notable development issue when recognising that affluent 'over-consumption' is not only increasing emissions but is damaging both societal and individual wellbeing (Fleurbaey et al. 2014; Bartolini 2014). Economic and structural measures to move towards services and the technological measures to improve efficiency are indeed necessary, but they are not sufficient. As more and more of the 'low-hanging fruit' is picked off through technological change, these development issues will inevitably surface and present a greater challenge to EU sustainability and climate policy. However, it has been posited there are many synergy or win-win outcomes available that could improve the quality of development and of people's lives. The EU has been a pioneer of mitigation but this continued status is in jeopardy if it does not increase the limited ambition of its reduction targets. According to the climatetracker.org assessment of 2016, current EU plans are not continuing the historical trend of reducing emissions, and increased action is needed to meet longer term targets. This necessitates not only accelerating the *technological transition* but adopting a *sustainable development pathway* which places social wellbeing and environmental sustainability and not economic growth at its core. Threats to EU policy include the disruption from the proposed British exit from the EU ('Brexit'), lasting effects from the 2008 economic recession including austerity policy, inadequate sustainability policy and support for renewable energy, and future demographic imbalances. The austerity policies implemented to address the recession have seen the social project of the EU take a backseat to the economic. The conversion of private into public debt, increased inequality and declines in public services have driven social discord and declining trust in public institutions, and could even be seen as the chief impetus towards Brexit. Reactionary voices continue to seek the wrong answers to the wrong problems, targeting immigration and multiculturalism as the source of ills, in a triumph of rhetoric over reason.

However, the EU has a number of key strengths. It has the economic and political capacity to implement effective policy, a strong and progressive social ethic in many countries, an international and cooperative approach to problem solving on climate, and a willingness to regulate and implement policy where evidently necessary. It has also tended to transfer the drive for sustainability and environmental protection from its more concerned members such as the Nordics (Box 6.1), to its less concerned states. While it is further along the sustainability transition, to ensure its success and realise development win-wins, it needs to fully harness integrated development and move beyond the technological transition, to more mainstreamed sustainable development, with the interrelated wellbeing of society and environment at its core.

Box 6.1: The Nordics: Seeds of Change Among High Emitters

The Nordic countries comprise Denmark, Norway, Sweden, Finland and Iceland; apart from Norway and Iceland, they are all full members of the EU and are often held up as progressive global development leaders. They consistently rank as 'best-in-class' in development metrics including happiness, healthy life years, income and equality, the social progress index, the human development index, environmental quality, corruption and a host of others. But why such enviable results in the Nordics in comparison to inferior outcomes in some of the wealthiest countries such as the US?

The progressive social democratic political models of the Nordics involve higher taxes and a related strong investment in public services, a priority on environmental protection and the protection of freedoms, opportunity and human rights. The basic approach does not just come down to levels of public investment, or indeed to ideas of 'economic efficiency', but the priority placed on positive social and environmental outcomes in politics, and in wider society. While they are free market economies, they are influenced by a strong social ethos which guides appropriate state intervention across society. This contrasts with the social, economic and environmental failures in more laissez-faire market economies worldwide. The influence of the Nordic countries has also aided the drive towards social and environmental protection throughout the EU.

Most of the Nordics have had the advantage of starting from an early deployment of renewables, but their ecological footprint and GHG per capita remain relatively high in global terms. Much more needs to be done to achieve sustainable development pathways and low-carbon transition. Nevertheless, the progressive approach to both social and environmental outcomes, leaves them well-placed to lead, both in achieving climate protection and in promoting the wellbeing of citizens and the environment into the future.

The US, the Potential to Become a Major Player?

The pathway of US involvement in sustainability and mitigation embodies many contradictions, particularly in the last two decades. As an historical pioneer of conservation and environmental protection, and one of the chief architects of the Kyoto Protocol, it was a major supporter of 'market mechanisms' such as emissions trading, but it was also effectively the first country to shun its ratification. A Senate vote in 1997 several months prior to agreement on Kyoto, rejected by 95-0, any treaty which facilitated the exclusion of developing countries from binding emissions targets.[4] The US repudiated Kyoto in 2001 but despite years of federal inaction, many states and cities implemented emissions targets, emissions trading and renewable and energy efficiency initiatives. However, rejection of what is seen as 'big government' remains a major ideological issue, despite much evidence of the benefits of appropriate state intervention. The US has a strong tradition in scientific research, particularly in the science of climate change and also in technology innovation, and a long political heritage of empowerment through civil society and environmental NGOs. Nevertheless, it has also been the origin of a concerted campaign to weaken climate action through denial of the scientific evidence funded by vested interests[5] and corporate lobbying.

While sub-national efforts improved the position of the US, the lack of sufficient domestic action not only meant that national emissions did not meet Kyoto targets, but the US became a key stumbling block to reaching agreement with developing countries for the 'post-Kyoto' period. Yet the Obama administration (2009–2017) helped ramp up domestic action on climate change, and engaged constructively in international agreements under the UNFCCC. The period prior to Obama could be described as 'lost decades' as 'carbon lock-in' deepened. As a result, it not only has higher environmental and social costs, but also increased economic costs: (i) the overall economy is more energy and resource intensive, (ii) opportunities for low-carbon growth are lost to competitors, (iii) there are increases in un-monetised social and environmental costs such as those from air pollution, (iv) future transition costs are increased and (v) inaction raises the risk of future financial liabilities to those impacted by climate change. Continuing in this way is evidently a 'lose-lose' approach to development.

Current assessments of the US INDC, such as that by Greenblatt and Wei (2016), suggest that it may be possible to reach its long-term pledge to reduce GHG emissions by 83% on 2005 by 2050. This would have required a technological transition through full implementation of its 'Climate Action Plan' and its supporting 'Clean Power Plan', but the current administration seeks to halt these. A number of other distinct challenges are evident including high per capita consumption levels, high per capita energy consumption and emissions, infrastructure deficits, urban sprawl, social inequalities, citizen disenfranchisement, policy deadlocks and public debt. The patterns of development may have delivered increased income per capita, but the distributional picture is different, and the US is far from a sustainable development pathway. They have increased inequality without improvements in wellbeing, such as life satisfaction (see Table 6.1), and are undermining the social and natural capital on which the country is built (Bartolini 2014). There is great uncertainty in US climate policy, and it is optimistic to consider that plans developed under the Obama would be sufficient to reach long-term targets. A re-commitment to an effective climate policy, and a transition to a sustainable development pathway, is now urgently required. However, the US has characteristic strengths to leverage research and technological innovation and has economic, policy and civil society capacity for transformation. These are strengths which are being undermined by the actions of the Trump administration. Its historic responsibility for GHG emissions, and its global economic and diplomatic reach, places a great responsibility on the US in climate action. There is also a great opportunity for win-win outcomes if it responds effectively, and if it changes the political narrative to a rational debate on outcomes that benefit all.

Japan: From the Limits of Technology to a State of Flux

Japan, the sixth largest GHG emitter in the world,[6] was in the vanguard of climate action at the time of the Kyoto Protocol, taking on a target of a −6% reduction in GHG emissions during the Kyoto period of 2008 to 2012. One single significant event, the Fukushima Daiichi nuclear accident in 2011, was a watershed moment in Japanese climate policy, with all of its remaining 48 working nuclear reactors going offline due to safety fears. GHG emissions had decreased as the recession emerged in 2008, but rapidly increased once more. Reduced nuclear power generation was replaced by coal, oil and gas. It is now the only G7 country seeking to significantly expand electricity generation from coal, and has plans for at least 43 new plants in the next 12 years. As it had breached its target, Japan complied with Kyoto by using what are known as 'flexibility mechanisms' in the language of international climate politics. It purchased credits for emissions that were reduced overseas, and used accounting of carbon emission reductions arising through natural carbon sinks in Japan, known as 'Land use, Land-use Change and Forestry', or 'LULUCF'.[7]

The early industrialisation of Japan was accompanied by a string of environmental catastrophes that began in the nineteenth century. Copper poisoning from the Ashio mine in 1878, was followed by the discovery of '*itai-itai*' and '*minamata disease*', from the 1950s onwards, with poisoning over decades resulting from releases of toxic cadmium and methylmercury. The period of post-war reconstruction and economic growth was also accompanied by problems of air pollution and water quality. Environmental policy and legislation were successively strengthened in response, but this lag period in responding sufficiently to environmental challenges accompanying economic development may have a repetitive quality. While Japan's technocentric response showed some success with respect to air pollution, current plans for coal appear a major regression. Japan is known internationally as a technological leader, which could support the technological transition. Its comparatively low GHG emissions per capita and ecological footprint (Table 6.1) may favour a sustainability transition. There have been moves in the right direction with a carbon tax implemented in 2012, and various improved sectoral measures such as renewable energy promotion and vehicle efficiency standards, as discussed by the World Resources Institute assessment of 2014 (Kuramochi 2014). Indeed Japan has a long tradition of effective policies for energy efficiency. According to a recent assessment by the International Energy Agency (IEA), these partly explain why total primary energy supply (TPES), total final consumption of energy (TFC) and electricity demand all peaked in the last decade (OECD/IEA 2016). Its INDC, submitted before the Paris climate summit, was to reduce GHG emissions by 26% from 2013 to 2030. In May 2016, it adopted the 'Plan for Global Warming Countermeasures' pledging to seek cuts in emissions of −80% by 2050, conditional on compatibility with economic growth. This qualifying condition could encourage the country to seek cost-effective technological transition, or it could be a get-out-of-jail-card to avoid sufficient short-term action.

Noting the economic benefits of action, particularly for a technological leader, the warning of history is prescient in Japan's case. The government is working in partnership with industry and academia to promote energy technology innovation under the 2016 National Energy and Environment Strategy for Technological Innovation towards 2050 (NESTI 2050). The IEA notes three critical areas to success: energy efficiency, increasing renewable energy supply and restarting nuclear power generation. However, it is clear that such a technological focus does not address a sustainable development pathway. It may be that the 2030 target is within reach, however it is not compatible with the ambition required to meet a +2°C target globally according the Climate Tracker 2016 assessment. Excluding forestry and land use it may involve only a 4–11% reduction on 1990 by the year 2030, or in the ballpark of the original Kyoto target 20 years after its expiration. The longer term path to 2050 is inadequate even according to the upbeat assessment of the IEA (2016). The greening of government programme implemented in the 1990s, the Low-carbon City Act (2012) and the 'Lo-House' promotion of more sustainable housing and lifestyles (2006) offer some sign of a move towards a sustainable development pathway. However, there is still a heavy reliance on sectoral, technical and individual behaviour measures, contrary to what is known about the limits of such sustainable consumption and production approaches (Fleurbaey et al. 2014). Japan has faced other challenges in recent decades. The high GDP growth rates of the 1980s gave way to stagnation in the 1990s, and a deep recession beginning in 2008 placed obstacles to the necessary investment required for technological transition. Social inequality and poverty increased in Japan in the 1990s, attributed to demographics, low levels of social spending and increasing 'non-regular workers'[8] who have lower benefits (OECD 2010a).

Apart from its early efforts to implement environmental policy, Japan has a history of action on energy and emissions. Its response to the oil crises of the 1970s, led to the 'Sunshine Project' in 1974 and the 'Moon Light Project' in 1978. This enabled a 30% improvement in energy consumption efficiency over 20 years and enhanced research in renewables and advanced technologies. The effects of the nuclear shutdown on emissions plans have undoubtedly created a significant headache for lawmakers. Meeting emissions reduction commitments will mean harnessing technological ingenuity, significantly increasing targets in the coming years and, crucially, enabling a sustainable development pathway in addition to technological change. The ancient East Asian concept of *Mottainai*, conveying a sense of regret concerning waste, may have relevance in moving from economic and technological hegemony, to a new age of understanding the need for integrated thinking and a balanced approach to development.

The 'Emerging Market' Giants: China, India, Brazil and Mexico

Though officially still classed as 'developing countries', China, India, Brazil and Mexico have, since the mid-2000s begun to offer leadership on climate change in conjunction with their emergence as fast-developing globalising economies. While China's average annual GDP grew by 10.3% between 2000 and 2014 and India's grew by 7.5% over the same period, the two Latin American giants grew less over the same period, Brazil by 3.7% and Mexico by 2.3%. Selected development indicators for these countries are included in Table 6.1. In 2015, Brazil entered into a severe economic downturn that led to growing political unrest and the impeachment of President Dilma Rousseff in mid-2016. Yet, unlike other large 'emerging market' economies like South Africa, Indonesia or the Russian Federation, or successful emerging economies like South Korea, Singapore, Chile and Argentina, the four emerging giants being profiled here are all classed by the Climate Action Tracker as showing 'medium' effort on curbing emissions. This puts them in the same category as the EU and the US, and above Japan, whose actions are classed as 'inadequate'. It marks a significant breakthrough for a group of countries which bear a lower historical responsibility for climate change. How, therefore, are these countries planning to reduce their emissions, and how adequate are their plans?

China: A Growing Awareness Taking Hold

China is famed for many defining characteristics, from its ancient culture and philosophy to its modern mix of hyper-capitalism and communism. Its sky-rocketing growth, through establishing itself as a global hub of manufacturing, has been aimed at raising the world's largest population from poverty, but results have been decidedly mixed. As evident in

Table 6.1, inequality is problematic and wellbeing metrics are not showing a notable benefit from rising incomes (Bartolini 2014). There are well documented concerns about labour issues and the protection of freedoms and rights in general. Environmental problems are mounting. The reliance on coal has caused serious issues with air pollution and public health, and a steep rise in GHG emissions has taken China to the top of the global emitters table. Urbanisation, poverty, water pollution, deforestation, desertification, water shortages and biodiversity loss, also number among its chief development challenges, and have driven social unrest and community activism. Noting these social and environmental outcomes, and the emergence of economic problems that are related to these patterns of economic growth,[9] it is clear that prioritising economic growth as the approach to development is proving a false friend in China. This is a classic example that supports the conclusions in the field of development economics that the focus must be on sustainable human development and not economic growth alone.[10]

China's commitment to action on climate change began to change significantly in the 2005–2007 period. The publication of its first National Climate Change Programme in 2007 was accompanied by commitments in its twelfth five-year plan in 2011, to increase the share of non-fossil fuels in primary energy consumption, and, to compulsory reductions in carbon intensity and energy intensity.[11] In 2011, the Communist Party Central Committee approved proposals to establish an emissions trading scheme. Though China announced its first national targets to limit the growth of GHG emissions just before the UN CoP in Copenhagen in 2009, it resisted establishing international targets at that conference. However, two years later at the Durban summit in 2011, it agreed to the process to negotiate a global treaty that would include targets for all countries, not just developed countries, as was the case under the Kyoto Protocol. These marked important changes in China's position. In November 2014, President Xi Jinping and President Obama together announced their countries' post-2020 climate change targets, with China pledging to peak its GHG emissions 'around 2030', the first time it had agreed such a specific target.

Its INDC, submitted prior to the Paris CoP in 2015 reflected these developments. It repeated its commitment to peak carbon emissions by 2030 at the latest, to lower the carbon intensity of GDP by 60%

to 65% below 2005 levels by 2030, and to increase the share of non-fossil energy in total energy supply to around 20% by the same date. It also committed to increase its forest coverage by 4.5 billion cubic metres compared to 2005 levels.[12] While acknowledging the steps taken to develop renewable energy, including an increase of 400% in solar energy capacity since 2005, and actions to tackle non-CO_2 emissions, the Climate Action Tracker identifies 'large uncertainties' associated with its targets, not least levels of future GDP growth. It points out that non-energy-related emissions are highly uncertain and are not covered by its pledges, while its carbon intensity targets seem to contradict other policy pledges, not least in regard to pollution. 'Unlike many governments, China's current policy projections embed many climate policy related actions and goals, and bring the country close to achieving its target for 2030', states Climate Action Tracker, while adding that its carbon intensity target is inadequate. Luukkanen et al. (2015) showed structural change in the economy away from manufacturing and suggest that the carbon and energy intensity targets require concrete policy steps to deliver by 2030.

As discussed by Luukkanen et al., China has taken steps in its 12th five-year plan to transform from an investment and export-driven economy to an economy driven mostly by domestic consumption. It has also taken the unique step of establishing exact GDP growth targets that, at least in theory, should not be exceeded. Its National Report on Sustainable Development (PRC 2012: 3) aims to put people first as its core, as a 'scientific approach to development' that is balanced and sustainable, in harmony with nature, while protecting rights and social progress. If this indicates growing awareness among lawmakers, citizens and the market in China, then it may drive not just a technological transition, but a sustainability transition in general towards human and environmental wellbeing. It is also notable that they seek to use a 'scientific' evidence-based approach to policy and development. Many questions remain about how effectively this will be implemented, and what kind of outcomes it will generate, particularly when noting lax environmental regulation in the past. China's position as the world's largest GHG emitter since 2006 means that how this occurs has global implications. China's determination to once more establish itself at the centre of global affairs will depend on its ability to embed a rapid correction of the imbalances that have arisen. Box 6.2 discusses how China is attempting to move to more balance by targeting a reduction in average meat consumption.

Box 6.2: Targeting Meat Consumption in China

In mid-2016, China's health ministry announced plans to reduce its citizens' meat consumption. The average Chinese person currently eats 63 kg of meat a year, and this is expected to increase to almost 100 kg by 2030. The guidelines would reduce this to between 14 kg and 27 kg a year. This would lower emissions from the livestock industry by 2030 from 1.8bn tons to 0.8bn tons. As well as cutting GHGs in the atmosphere, the move would counter strains on water supply and on arable land and help address rising levels of obesity and diabetes. China currently consumes 28% of the world's meat and this consumption is estimated to add 233 million tons of GHGs to the atmosphere each year.

Analysts point out, however, that Chinese companies are buying up farms in the US and Australia to help meet demand for meat at home. Jeremy Haft, an adjunct professor at Georgetown University in Washington DC, says 'China's consumption of meat is skyrocketing. ... From a climate perspective, the methane will still be created, but will be shifted to the United States' (Milman and Leavenworth 2016). Furthermore, policies of European agricultural exporters such as Ireland and France are scrambling to expand the Chinese market for beef and dairy, despite the significant problems it creates. Such artificially inflated markets drive economically inefficient forms of food production, at the expense of public health, the environment and food security, while livestock agriculture accounts for about 15% of global GHG emissions.

China's Health Ministry plans to reduce meat consumption may be a signal of a more rational direction in global food and agriculture policy. The increasing public concern towards physical health could also begin to weaken the global aspiration for increased meat consumption, as a relic of twentieth century ill-health. According to World Resources Institute in its *CAIT emissions data explorer*, in 2009, the average person in more than 90% of the world's countries consumed more protein than necessary to meet average daily requirements.

India: Evolving as Global Sustainability Leader?

India is a country with a rich natural endowment and a cultural and philosophical heritage that spans millennia of human advancement. Its immense population will soon surpass China's as the world's largest, but the country has significant problems of extreme poverty, low levels of human wellbeing and environmental degradation. It has traditionally been resistant to taking legal responsibility for climate change, seeing it as primarily a problem for developed countries and instead giving priority to national development, consistent with its low per capita emissions and its low historical responsibility for climate change. However, signalling a new willingness to proactively address the issue, a Prime Minister's Council on Climate Change was established in 2007 and a National Action Plan on Climate Change published the following year. This established eight 'national missions' for the period to 2017, including establishing India as a global leader on solar energy and enhancing energy efficiency, reducing consumption and carbon emissions. While setting ambitious goals for economic growth, the 2012 five-year plan also sought to reduce the emissions intensity of that growth by 20% to 25% by 2020. However, this excludes emissions from agriculture. Although India has strongly resisted any targets for reducing emissions, it did support the Durban Platform of 2012 to negotiate a climate treaty applying to all countries. Despite these advances, unlike China, India has refused to commit to a date for peaking emissions.

India's INDC, submitted just before the Paris summit, commits to reducing the emissions intensity of GDP by 33% to 35% below 2005 levels by 2030, increase the share of non-fossil energy to 40% of installed electric power by 2030 (though this seems dependent on technology transfer and international funding from the Green Climate Fund) and increased carbon sinks through additional forest and tree cover by 2030. However, as the Climate Action Tracker assessment points out, though the growth rate of solar and wind-power capacity is greater than for coal-powered electricity capacity, there will be an absolute growth in demand for electricity. This means that growth in coal-powered electricity generating capacity will actually outstrip the total increase in the generating capacity from renewables. Overall, based on current policies, the assessment expects India's per capita emissions to increase by around 84–88% by 2030, though this would still leave them below the world average in 2010. The assessment concludes that India's climate plans 'are at the least ambitious end of what would be a fair contribution'.

India's low ecological footprint and GHG emissions per capita (Table 6.1) mean that it has a great opportunity to pursue a sustainable development transition if it promotes a balanced development. Its reach extends to meeting rapidly growing energy demand through seeking to develop hydropower in the Nepalese Himalayas. Such developments pose risks of damaging local ecosystems and human rights without careful assessment and integrated thinking. Rapid growth in renewable energy, reductions in coal imports and the cancelling of large coal projects suggest that the technological transition may have begun. As stated by Climate Action Tracker, these developments 'are amongst the most important underway globally'. The *Deep Decarbonisation Pathways* project for India (Shukla et al. 2015) has established that it is feasible to decarbonise India's economy with current technologies, and with significant co-benefits for pressing public health and environmental issues such as air pollution. The pathways from this study are very useful for characterising the energy and technological transition, but there is an urgent need to branch into the more fundamental challenge of how to achieve sustainable development. The partnership that led to the *Low Carbon Development Pathways for a Sustainable India* study (Parikh et al. 2014), point out that combining transition with increasing investment in goals such as education and health, can lead to poverty eradication, sustainable development and the attainment of threshold values for human wellbeing sooner. The partnership concluded that the visionary development pathway does not involve any significant cost compared to a business as usual scenario, and has the potential to play a critical role in advancing political and public discourses on integrated climate change mitigation and development in India. While more research is required, this groundbreaking integrated approach to transition is an important step. Visions of sustainable development pathways for India are among the most important tasks for human development, climate change and the environment. It is unlikely that the vast swathes of humanity that make up India's population can achieve acceptable levels of development, without both global action to arrest climate change, and local action to achieve sustainable development.

Brazil: Balancing Development in a Climate Leader

Brazil is a famously multicultural nation, and is home to the Amazon rainforest with the immense biodiversity that it holds. Climate change is of great concern to this jewel of Brazilian natural wealth, as increasing temperatures will threaten the future viability of the 'lungs of the earth'. As part of the negotiations on the Kyoto Protocol, the Brazilian delegation proposed in May 1997, to set differentiated emission reduction targets for countries according to the impact of their historic emissions on temperature rise. The 'Brazilian proposal' was a visionary approach to the ethics of development in a climate-constrained world.

Brazil's relatively low GHG per capita emissions can be deceptive (Table 6.1). When emissions from land use change and forestry (LULUCF) are included, they almost double, from 5.1 to 9.18 tonnes per capita 2012.[13] This is an illustration of the importance of the wider context of emissions, and 'sustainable development' rather than just energy when addressing greenhouse gas emissions. Brazil was a global pioneer in biofuels, developing ethanol fuel from sugar cane. Following the oil crisis in 1973, it initiated the National Alcohol Programme in 1975 (ethanol is derived from the alcohol in sugar cane), with the goal of phasing out entirely the use of fossil fuels for cars. Since 1976 it has been mandatory to blend ethanol with oil products as a transport fuel; since 2007, ethanol must constitute at least 25% of such fuels. Unlike other biofuels, ethanol does not displace food production and has no destructive impacts on ecosystem services. The switch to ethanol has required making flex technology for vehicle engines generally available. As a result, 'the Brazilian automobile industry has come to be considered as part of the green economy' (Abramovay 2016: 159). Brazil also relies heavily on hydroelectric plants to generate electricity with as much as 81% of its electricity being generated by this means in 2011 though the percentage has since dropped. Another 30 such plants are planned in the Amazon region by 2023 (ibid.: 156). Brazil can therefore be said to be in a strong situation to reduce its energy-related GHG emissions. It was one of the first major developing countries to pledge an emissions reduction target when it announced in January 2010 that it would reduce emissions by between 36% and 38% by 2020. This was enshrined in law in December of the same year, and was not conditional on international funding as are the targets pledged by many developing countries.

The INDC it submitted before the Paris summit committed to reduce net GHG emissions by 37% below 2005 levels by 2025 with an 'indicative contribution' of a 43% reduction by 2030. This includes a share of 45% of renewables in the total energy mix by 2030 (currently at 41.3%). Already Brazil has reduced its emissions by 41% between 2005 and 2012, mainly due to significant reductions in forestry loss and land use change. With about

60% of the Amazon rainforest in its territory, extensive deforestation had led to very high emissions in the late 1990s and early 2000s. Strong public policies had reversed this trend and emissions from deforestation had fallen by 85% below their highest level, reached in 1996. However, with a big increase in demand for energy, the government has been auctioning permits for coal and gas-fired power plants to increase the flexibility of the power sector and overcome the volatility associated with a dependence on hydroelectric generation. The Climate Action Tracker assesses that this policy 'may ultimately limit the options for deep decarbonisation into the more distant future'. The country also has ambitious plans to reduce illegal deforestation to zero between 2025 and 2030, to reforest 12 million hectares and to restore an additional 15 million hectares of degraded pasturelands by 2030, thereby strengthening carbon sinks. Abramovay cautions that the expansion of mining coupled with investments in large-scale hydroelectric projects in the Amazon region may interfere with achieving these goals (Abramovay 2016: 163). Overall, the assessment concludes that Brazil could strengthen its emissions reduction target to reflect its 'potential to increase energy efficiency and develop renewable sources of energy'.

Despite its relatively strong showing in measurements of life satisfaction, Brazil has a significant challenge of social inequality (see Table 6.1). This is linked to problems of social unrest and high crime rates. A redistribution of the benefits of economic development could therefore be identified as a priority action in social sustainability for Brazil. A sustainable development pathway will also necessarily involve: rainforest conservation, protection of biodiversity and the rights of indigenous people, addressing land degradation from mining, and air and water pollution. An integrated approach to development would balance social quality with these environmental challenges.

Mexico: Climate Policy Needs to be Matched by Social Priorities

Mexico was home to advanced ancient cultures such as the Aztecs, Maya and Olmec peoples. It is regarded as a vibrant democracy with a young population, and has the 15th largest global economy by GDP. It was the first developing country to set a firm emissions reduction target for 2050. According to a study prepared by Ecofys and Climate Analytics, Mexico 'has made some of the fastest advances of any country in the world in strategic planning on how to incorporate low carbon development into all parts of the economy' (Ecofys and Climate Analytics 2012: 4). This was due to the strong personal commitment of President Felipe Calderón (President of Mexico 2006–2012) and followed on from the establishment of the Inter-Ministerial Climate Change Commission in 2005 to co-ordinate strategic planning across government. Mexico's planning and institution building to address climate change has been described as 'remarkable' as it was based on a high-level of awareness penetrating a wide circle of stakeholders and actors, the availability of extensive data, and clear institutional responsibilities, lines of communication and focal points to ensure consistency and policy development (ibid: 4).

One of the first of its kind in the world, the General Law on Climate Change adopted in 2012 included a pledge to reduce emissions by 30% by 2020 and by 50% by 2050, conditional on international financial support. Its current four-year plan up to 2018 includes 28 mitigation measures and an overall strategy for the 50% emissions reductions by 2050. Mexico has also taken an active role in international climate diplomacy: then foreign minister Patricia Espinosa Cantellano's chairing of the Cancun climate summit in 2010 was widely credited with putting the process back-on-track to negotiate a successor to the Kyoto Protocol. She became executive secretary of the UNFCCC in 2016.

In its INDC for the 2015 Paris summit, Mexico pledged to reduce emissions by 22% by 2030 unconditionally, and by 36% if certain conditions were met, including access to low-cost financial resources and technology transfer, as well as a global agreement addressing the international carbon price. Mexico's pledge includes the specification that it be economy-wide and includes a comprehensive accounting of all sources and gases, including land-use change and forestry. Indeed, emissions from agriculture, land-use and deforestation have been declining since the 1990s, though over the same period energy emissions have been increasing substantially. Mexico also specifies a 'net emissions peak starting from 2026'. Climate Action Tracker ranks Mexico's targets as 'medium' but says it would upgrade them to 'sufficient' if it were to adopt its more ambitious 2030 targets without preconditions.

Mexico has developed on a path of low tax and low investment in public services (OECD 2010b), and has associated problems of social inequality and poverty, with limited social protection to offset this. Access to tertiary education is limited for the less affluent according to the OECD, and this will further compound lack of opportunity, crime, lower health and well-being, thus weakening Mexican society. A groundbreaking study on the health-related outcomes of the UN Sustainable Development Goals

(SDGs), published in the *Lancet* by Lim et al. (2016), ranked Mexico at 69 out of 188 countries, noting problems with violence, disasters and unsafe water. Reforming subsidies for fossil fuels and increasing the carbon tax could be useful tools to achieve this re-balancing, when accompanied with social welfare measures to protect the less affluent from higher costs. The commitment to low-carbon development holds potential to improve public health and reduce emissions in Mexico. It needs to be matched by stronger political commitment to prioritise the social dimension of development if sustainability and the wellbeing of its citizens are to be achieved.

Conclusions

This chapter looked at the pathways being taken in a diverse range of economically advanced and emerging countries. The developed countries were separated by common but differentiated responsibilities under UNFCCC rules, but as the challenges and opportunities have crystallised since the 1990s, this differentiation has weakened. The intention has moved from short-term mitigation in the developed countries to long-term transition in all countries, albeit at different rates.

While the EU has achieved well on many measures (Table 6.1), and shown climate leadership since the 1990s, there are now challenges to this leadership role. It needs to establish a wider sustainability focus as a development path approach to climate policy, to address its significant material consumption, to deepen targets and to accelerate the technological transition. It also needs to address the social impacts that have emerged from recession and austerity. It is well placed to evolve on a sustainability path that presents many opportunities but it needs to respond in a more integrated and effective fashion to avoid reactionary positions that weaken its potential to achieve win-wins. The US evolved from climate laggard to climate leader under the Obama administration, playing catch-up on its European counterparts. However, it has many sustainable development challenges with poor social and environmental outcomes (Table 6.1). It is not evolving on a sustainability path and the climate leadership shown under Obama is now being dismantled by President Trump, with the sidelining of the Climate Action Plan. The market-led approach in the US does not appear to be delivering on any metric apart from enrichment of a tiny minority, who have continued benefitting even during the recession. While the US has much potential, this Trump administration policy approach suggests a deep regression. Japan is another example of the limitations imposed by giving priority to the technological transition, accompanied by the retrograde step of the turn towards coal. It presents a number of constructive outcomes but is not evolving on a sustainability path. Similar to the EU, it needs a move towards integrated thinking and balanced development, but in contrast to Europe it requires an even stronger focus on the technological transition.

The emerging countries show a commonality in that they are all moving towards accelerated climate action, in contrast to the US. They have all shown some economic success, nevertheless the many social challenges place a question mark over their development models. Brazil and Mexico need to improve distribution of income and public services. China and India also need to address inequality, public services and environmental quality (Table 6.1), but show some initial signs of moving towards sustainable development pathways. India's low ecological footprint is exemplary, but not if it is retained through continuing poverty. Inequality is emerging to a greater or lesser extent across all countries considered. While different development models are required reflecting diverse circumstances, all of the examples considered suggest the emergence of limitations. The techno-economic approaches appear necessary but insufficient. There is a need for sustainable development pathways and integrated development policy as the fundamental approach to enable transition. This was a conclusion offered by the IPCC as a 'development first approach' in 2007 (Sathaye et al. 2007: 695). Such pathways can also be directed to deliver the most desirable social outcomes, robust economies and environmental integrity. While recognising that there may be trade-offs, this synergistic approach to policy is the key to pursuing the many opportunities and achieving win-wins.

This chapter moves the focus from the technological transition to the political economy of how low-carbon transition and the sustainable society are achieved. It is clear that moving to a sustainable development pathway rather than just techno-economic measures presents greater potential to reduce emissions and also leads to better development outcomes. Sustainable development is used as a framework for thinking but one requiring clear policy definition in each country. Various political economy approaches are evident across the countries surveyed. All have experienced economic successes, and some have combined this with progress on mitigation, environmental protection and social sustainability. While the EU has achieved more success in developing a sustainable society than the US, it may be backsliding, as market needs appear to be taking precedence over social and environmental objectives. There are also similarities in Japan. The US has strongly grown its economy, but regressed environmentally and on many social measures, where the free market is even stronger in its dominance. The emerging countries also all appear to have prioritised the market at the expense of social and environmental outcomes. Nevertheless, the state continues to give direction, particularly in China, and each country shows evidence of some progress.

While state intervention through public policy is contested (Shafaeddin 2004), the examples discussed appear to show that intervention is necessary, both for low-carbon transition and for a sustainable society. The outcomes from the more neoliberal approaches are contrary to the interests of transition[14] and strongly counter to a just and sustainable society (Richardson et al. 2016; Bartolini 2014). The positive outcomes in the social democracies of the Nordic countries are worth noting in that they have evolved as capitalist market economies but retained a strong focus on social and environmental outcomes. Appearing as the most successful development examples of the countries considered, they embody an approach in which the state endeavours to give priority to the interests of citizens and social development over the interests of capital and economic growth. The low-carbon transition necessitates not only accelerating the *technological transition* but adopting a ***sustainable development pathway*** which places social wellbeing and environmental sustainability and not economic growth at its core. China's use of economic growth targets that are not to be exceeded is a novel approach. Political economy therefore provides a useful lens on how this process can be achieved in practice. Chap. 7 examines a range of different routes to development and sustainability, from Latin America and Africa, to Asia and the Small Island Developing states, concluding by distilling the lessons for political economy across the two chapters.

Notes

1. The Kyoto Protocol applies to 192 countries and places various responsibilities on all parties. However, quantified emissions limitation and reduction objectives (QUELROs) that didn't apply to the non-Annex I, or less developed parties, reflected not only CBDR, but the absence of a political consensus on what avoiding 'dangerous anthropogenic interference with the climate' actually involves. This was required to establish the global carbon budget and emissions trajectories necessary to meet the defined climate objective.
2. The Ecological Footprint measures humanity's demands on nature in terms of 'the area of land and water it takes for a human population to generate the renewable resources it consumes and to absorb the corresponding waste

it generates' (footprintnetwork.org). It is thus a wider measure than the carbon footprint which measures the amount of carbon and other GHG gases emitted. The EF approach includes a measure of how many Earths would be required to support each national Ecological Footprint (see column six of Table 6.1).

3. Including the environmental impacts of mining, drilling, transporting and processing of coal, oil, gas and peat on water, air, land and habitats. Air quality is damaged by air pollution (Sulphur dioxide (SO_2), Nitrogen oxides (NO_x), Particulate Matter (soot), mercury and radioactive substances) when fossil fuels are burned, leading to major public health impacts. The impacts of climate change arising from the burning of fossil fuels are significant: damaging economic sectors, impacting human life and wellbeing and further weakening environmental integrity. There are also long-term risks of environmental and social collapse arising from climate change.
4. The Byrd-Hagel resolution also stipulated that the treaty should be rejected if it would 'result in serious harm to the economy of the United States' something repeatedly rejected by economic analysis as costs would be 'insignificant' and potentially even of net benefit (Barker and Ekins 2004). It therefore appears more as a political strategy as it was not based on robust evidence.
5. As discussed by the scientific historians Oreskes and Conway (2012), the same individuals posing as experts, and using the same techniques, have been active in the denial of climate change science, as in denial of the links between tobacco smoking and lung cancer. These 'merchants of doubt' achieved some success in 'keeping the controversy alive' by spreading doubt and confusion. The science was fudged by misrepresenting confidence and uncertainty to confuse public opinion, as was also evident in the now infamous RICO fraud of the tobacco industry in previous decades (United States vs. Philip Morris).
6. World Resources Institute CAIT emissions by country 2012 excluding LULUCF.
7. Where carbon emissions are absorbed by natural processes in the growth of forestry and vegetation, and in soils.
8. 'Non-regular workers' includes part-time workers and other precarious employment groups such as dispatched workers, contract workers and temporary employees. While some argue that such arrangements contribute to economic competitiveness through employee cost reductions, the social cost is significant. These arrangements reduce incomes, increase inequality and poverty, and risk social hardship and unrest. It is widely known that such inequality leads both to reduced individual and societal wellbeing (Fleurbaey et al. 2014: 311). To avoid a race to the bottom, a broad or 'global' reconsideration of this spreading false economy is urgently needed.

9. That also includes debt, bubbles and unequal growth.
10. Anand and Sen (2000) have argued that the preoccupation with commodity production, opulence and financial success has taken the focus away from the more useful and important goal of human development. The lessons from development studies identify the economy as the *means* and not the *end* of development.
11. To reduce carbon emissions and energy consumption per unit of economic output.
12. Increasing the rate of removal of carbon emissions from the atmosphere through this natural 'carbon sink', with the removal then included in national emissions accounting.
13. Emissions per capita from WRI CAIT climate data explorer.
14. As they are delivering efficiencies but not absolute reductions in emissions (Fleurbaey et al. 2014), and are not yet toppling the incumbent fossil fuel technologies at the rate that is compatible with meeting the +2°C global climate target.

References

Abramovay, Ricardo. 2016. The Green Growth Trap in Brazil. In *Green Growth: Ideology, Political Economy and the Alternatives*, ed. Gareth Dale, Manu V. Mathai, and Jose A. Puppim de Oliveira, 150–165. London: Zed Books.

Anand, S., and A. Sen. 2000. Human Development and Economic Sustainability. *World Development* 28 (12): 2029–2049.

Barker, T., and P. Ekins. 2004. The Costs of Kyoto for the US Economy. *The Energy Journal* 25: 53–71.

Bartolini, S. 2014. Building Sustainability through Greater Happiness. *The Economic and Labour Relations Review* 25 (4): 587–602.

Ecofys and Climate Analytics. 2012. *Climate Action Tracker: Mexico*. Accessed 24 August 2016. http://climateactiontracker.org/assets/publications/publications/WP1_MX_Country_report_2012_exec.pdf

EEA. 2015. *The European Environment State and Outlook 2015*. Synthesis Report, Copenhagen: European Environment Agency.

Fleurbaey, M., S. Kartha, S. Bolwig, Y.L. Chee, Y. Chen, E. Corbera, F. Lecocq, et al. 2014. Sustainable Development and Equity. In *Climate Change 2014: Mitigation of Climate Change. Contribution of Working Group III to the Fifth Assessment Report of the Intergovernmental Panel on Climate Change*, ed. O. Edenhofer, R. Pichs-Madruga, Y. Sokona, E. Farahani, S. Kadner, K. Seyboth, A. Adler, I. Baum, S. Brunner, P. Eickemeier, B. Kriemann, J. Savolainen, S. Schlömer, C. von Stechow, T. Zwickel, and J.C. Minx. Cambridge, UK and New York, NY: Cambridge University Press.

Greenblatt, J.B., and M. Wei. 2016. Assessment of the Climate Commitments and Additional Mitigation Policies of the United States. *Nature Climate Change*. Letter published online 26 September 2016. doi:10.1038/nclimate3125

Kuramochi, T. 2014. *GHG Mitigation in Japan: An Overview of the Current Policy Landscape*. Working Paper. Washington, DC: World Resources Institute. wri.org/publication/ghg-mitigation-policy-japan

Lim, S.S., K. Allen, Z.A. Bhutta, et al. 2016. Measuring the Health-Related Sustainable Development Goals in 188 Countries: A Baseline Analysis from the Global Burden of Disease Study 2015. *The Lancet* 388 (10053): 1813–1850.

Luukkanen, J., J. Panula-Ontto, J. Vehmas, J. Liyong, J. Kaivo-oja, L. Häyhä, and B. Auffermann. 2015. Structural Change in Chinese Economy: Impacts on Energy use and CO2 Emissions in the Period 2013–2030. *Technological Forecasting and Social Change* 94: 303–317.

Milman, Oliver, and Stuart Leavenworth. 2016. China's Plan to Cut Meat Consumption by 50% Cheered by Climate Campaigners. *The Guardian*, 20 June 2016.

OECD. 2010a. *Environmental Performance Reviews: Japan 2010*. Paris: Organisation for Economic Cooperation and Development.

———. 2010b. *OECD Perspectives: Mexico Key Policies for Sustainable Development*. Paris: Organisation for Economic Cooperation and Development.

OECD/IEA. 2016. *Energy Policies of IEA Countries: Japan 2016 Review*. Paris: Organisation for Economic Cooperation and Development/International Energy Agency.

Oreskes, N., and E. Conway. 2012. *The Merchants of Doubt*. London: Bloomsbury Press.

Parikh, J., K. Parikh, P.P. Ghosh, and G. Kedkar. 2014. *Low-Carbon Development: Pathways for a Sustainable India*. WWF-India/CEE/Laya/CASA/Welthungerhilfe/IRADE. Accessed 24 March 2017. http://awsassets.wwfindia.org/downloads/lcdp___report_low_res.pdf

PRC. 2012. *The People's Republic of China: National Report on Sustainable Development*. Report to the Rio +20 United Nations Conference on Sustainable Development held in Rio de Janeiro, Brazil, on June 20–22. Beijing, People's Republic of China.

Richardson, H.S., Erik Schokkaert, Stefano Bartolini, Geoffrey Brennan, Paula Casal, Matthew Clayton, Rahel Jaeggi, et al. 2016. Social Progress: A Compass. *International Report on Social Progress*, Chapter 2 draft for comment. Accessed 5 January 2017. https://comment.ipsp.org/chapter/chapter-2-social-progress-compass

Sathaye, J., A. Najam, C. Cocklin, T. Heller, F. Lecocq, J. Llanes-Regueiro, J. Pan, et al. 2007. Sustainable Development and Mitigation. In *Climate Change 2007: Mitigation. Contribution of Working Group III to the Fourth Assessment Report of the Intergovernmental Panel on Climate Change*, ed. B. Metz, O.R. Davidson, P.R. Bosch, R. Dave, and L.A. Meyer. Cambridge: Cambridge University Press.

Shafaeddin, S.M. 2004. *Who is the Master? Who is the Servant? Market or Government? An Alternative Approach: Towards a Coordination System.* Geneva: United Nations Conference on Trade and Development. Discussion Papers No. 175, August 2004. Accessed 24 March 2017. http://www.unctad.org/en/docs/osgdp20049_en.pdf

Shukla, P.R., S. Dhar, M. Pathak, D. Mahadevia, and A. Garg. 2015. *Pathways to Deep Decarbonization in India*. SDSN—IDDRI. Accessed 24 March 2017. deepdecarbonization.org/wp-content/uploads/2015/09/DDPP_IND.pdf

Torney, Diarmuid. 2015. *European Climate Leadership in Question: Policies toward China and India*. Cambridge, MA: The MIT Press.

CHAPTER 7

Development and Sustainability in the Global South: Different Routes to Transition and a Sustainable Society

The countries looked at in the previous chapter are already locked into high-carbon development trajectories. Of all the countries examined, only India has an ecological footprint that is below the level of sustainability but its development plans look set to move it beyond this quite quickly and it has not yet committed to a date when its emissions will peak. Throughout the rest of the world, there are many countries which fall well within the level of sustainability measured by their ecological footprint, since their average ecological footprint is less than the planet's biocapacity. Here the challenge is to find pathways to development, providing a better living standard for their citizens, that don't entail putting pressures on the biosphere that it cannot sustain. This chapter looks at different examples of developing or peripheral countries, in different regions, asking what pathways they are seeking to combine development with sustainability.

The chapter begins by examining a range of countries that seek to combine development and sustainability, drawing attention to the successful cases of Costa Rica and Uruguay which are then examined in greater detail. The following section looks at the Latin American region, and includes a particular focus on the role of environmental activism that is a noteworthy feature of environmental politics in that region. The third section turns to Africa, with a focus on the 'green state' and on some of the region's successful examples of combining development and sustainability. The following section on some Asian countries highlights the region's diversity,

P. Kirby, T. O'Mahony, *The Political Economy of the Low-Carbon Transition*, International Political Economy Series,
DOI 10.1007/978-3-319-62554-6_7

combining high levels of vulnerability with some highly developed countries pioneering 'green growth'. Before turning to analyse the political economy lessons to be learnt from these cases, the situation of the SIDS is examined as these are the countries whose very existence is threatened by climate change and whose diplomatic pressure has become a feature of global environmental diplomacy. The final section looks back over these two chapters on pathways to distil the lessons.

Before turning to examine the different cases, it is important to avoid treating them as if separate from one another. Pope Francis draws attention in his encyclical letter to the 'ecological debt' that the developed countries owe to the developing world:

> A true 'ecological debt' exists particularly between the global north and south, connected to commercial imbalances with effects on the environment, and the disproportionate use of natural resources by certain countries over long periods of time. ... The warming caused by huge consumption on the part of some rich countries has repercussions on the poorest areas of the world, especially Africa, where a rise in temperature, together with drought, has proved devastating for farming. ... In different ways, developing countries, where the most important reserves of the biosphere are found, continue to fuel the development of richer countries at the cost of their own present and future. (Pope Francis 2015: pars 51 and 52)

The challenges of finding pathways that combine sustainability and development is, therefore, a global problem that needs to take account of how the decisions made by powerful countries affect the life chances of those in distant parts of the world. This is at the heart of the ecological debt since it is the long legacy of colonialism that has configured the economies and even landscapes of many developing regions, and continues to do so through the demand for natural products. And, as the example of the SIDS shows dramatically, the fate of so many developing countries is to an extent outside their own control and depends on the actions of developed and other large countries to reduce their emissions. This, then, is the context in which a survey of developing countries and regions needs to be placed.

Combining Development and Sustainability?

As Lamb writes 'surprisingly little is known about historical low-carbon pathways of development. Which countries enable high levels of access to household energy services, education, nutrition, health and democratic rights, at levels of emissions far below the industrial average?' (Lamb 2016: 523). Yet, this is the essential challenge facing most of the world's countries. Finding pathways that allow countries combine both the improvement of their citizens' living standards and life chances, what we normally label development, while also reducing emissions, protecting biodiversity and living within the capacity of the planet's ecosystem, what we label environmental sustainability, is therefore of the utmost importance for the future of human life on this planet.[1]

The term 'low-carbon development' has emerged to define this challenge. But Urban and Nordensvärd remind us that 'low carbon development strategies vary according to different contexts and need to take into account national priorities and capabilities along with considerations of local knowledge' (Urban and Nordensvärd 2013: 219). Furthermore, they add, developing countries have the right to develop and to grow their economy thereby perhaps requiring a certain increase in emissions in the future. They place countries on a spectrum from the highly developed to the least developed. The high-income countries need to focus mainly on mitigation rather than on development, emerging economies focus on both mitigation and development (see the cases of China, India, Brazil and Mexico in Chap. 6), whereas the less developed countries tend to focus more on development and less on mitigation (see Box 7.5 on the Maldives). However, they stress that the core low-carbon development model needs to incorporate switching from fossil fuels to low-carbon energy, promote low-carbon technology innovation, protect and promote natural carbon sinks such as forests and wetlands, and formulate policies to support low-carbon practices and behaviours. They offer four contrasting approaches towards low-carbon development:

- Low-carbon growth: focus on production: growing the economy while reducing emissions through the use of clean technologies;
- Low-carbon lifestyles: focus on consumption: reducing emissions through changing consumption patterns;
- Equilibrium economy: focus on social development: adapt economy for social ends with neutral growth;
- Coexistence with nature: focus on sustainability: promote behavioural change within a low-growth economy.

While these distinctions help clarify different conceptual approaches, they tend to avoid the hard trade-offs that may be necessary and that are difficult for policy makers and political leaders to sell to citizens whose expectations are defined by images of mass consumption societies given saturation penetration by media. We therefore need to examine examples of countries that are attempting to combine development with sustainability. Lamb draws up a list of countries based on combined indicators of wellbeing (conditions for satisfying physical health outcomes; distributional issues; and 'personal autonomy' combining economic and social opportunities) and their emissions pathways (energy consumption and carbon intensity; rates of growth in energy consumption and carbon intensity; and timelines of mitigation including estimates of when emissions peak and begin to decline). In doing this he cautions that sustained rates of mitigation of around 5% a year 'have never been experienced outside of major economic shocks and downturns, and never on a consistent year to year basis, but they are in line with ... studies assessing transition pathways that avoid 2°C' (Lamb 2016: 526). This reminds us that the list of countries combining development with sustainability is based on projections derived from current trajectories rather than on outcomes actually achieved. Nevertheless, in the reality of today's world, this is the best that can be done.

Lamb ends up with a list of 20 developing countries that combine low emissions and high wellbeing that 'tend to be diverse in their climates, levels of trade, and population growth, but are constrained to low and middle incomes' (ibid.: 524). However, the number of countries begins to reduce across the range of wellbeing indicators. Thus, a wide range of countries scores highly on dimensions such as sanitation, electricity, nourishment and secondary school enrolment but only six meet three indicators (Albania, Armenia, Costa Rica, Egypt, Georgia and Uruguay) while only two meet four dimensions of human need (Armenia and Georgia). Correlating with energy consumption, all countries show initial low levels but diverge in terms of their trajectories to mid-century: Cambodia, Colombia, Guatemala, Honduras, Indonesia, Paraguay, Sri Lanka and Tunisia remain low energy due to extremely low growth rates; Albania, Armenia, Costa Rica, Ecuador, Morocco, Peru and Uruguay show higher rates of growth; while Brazil, Egypt, Georgia and Vietnam show extremely high rates of growth in energy consumption. This last group is expected to reach consumption levels matching those in the OECD today. However, a large number of countries have low-emissions trajectories, keeping them well within their per capita allocation required to keep warming below 2°C: Albania, Armenia, Colombia, Costa Rica, Ecuador, El Salvador, Guatemala, Honduras, Indonesia, Morocco, Paraguay, Sri Lanka, Tunisia and Uruguay. The emissions of most of these countries are expected to peak at levels below half of those of OECD countries (2 to 4 tonnes of CO_2 per capita against 10 tonnes per capita for the OECD).

While this expected low emissions trajectory is a positive finding, Lamb concludes that only four countries in his sample combine high achievement in wellbeing with low-carbon trajectories: Albania, Armenia, Costa Rica and Uruguay. Of these four, only Costa Rica and Uruguay score well on high levels of accountability and the rule of law. Of the 20 initial countries identified, therefore, only two can be said to combine a range of development indicators with a low-emissions trajectory. He concludes that there is a wide divergence evident within his group of countries: the results for many 'are in sharp contrast to the carbon intensive pathways followed by most nations in the global North, as well as recently emerging countries such as China', yet the countries in his sample also have 'diverse patterns of growth and stabilization, as well as persistent social and political challenges'. He wonders whether many have been able 'to satisfy a basic floor of health, energy and household services, but are not otherwise trapped in stagnating socio-economic conditions' (ibid.: 531–532). Focusing on the two countries that do exemplify the ability to combine development and sustainability, Lamb comments on Costa Rica that 'with both high physical and social need satisfaction, [it] speaks to the importance of a strong social democratic state in withstanding international market forces and pursuing human development progress on the periphery' (ibid.: 531). A similar point could be made about Uruguay (Box 7.1).

Box 7.1: Uruguay 'Defining Global Trends in Renewable Energy'

Among the many emission reduction pledges made before the Paris climate summit, that of Uruguay caught the attention: a pledge to reduce carbon emissions by 88% by 2017 relative to the average for 2009–2013. This it can do because 94.5% of its energy now comes from a range of renewable sources, and they account for 55% of the country's overall energy mix, including transport fuel. The global average is a mere 12%.

This transformation has happened recently. After the left-wing Frente Amplio reached power for the first time in 2005, a long-term plan was finally agreed after years of debate. In 2008 energy policy fixed a price for 20 years guaranteed by the state utility thus giving confidence to investors. Thus three elements are seen as being crucial to this success: clear decision-making, a supportive regulatory environment and a strong partnership between the public and private sector.

Furthermore, a mix of renewables including wind, biomass and solar ensures a resilience which means it has not had to import a single kilowatt hour of energy over recent years despite being dependent on electricity imports from Argentina in the past. The WWF named Uruguay among its green energy leaders saying that 'the country is defining global trends in renewable energy investment' (Watts 2015).

Yet, Uruguay faces the challenge of reducing emissions from its large herd of dairy cattle, four times the country's population of 3.4 million. Producing 80% of the country's methane emissions, Uruguay has pledged to cut this by a third through 'improving the efficiency of the emissions per product in the sector' (Pashley 2016).

The evidence of this section offers little support for the Environmental Kuznets Curve (EKC) hypothesis. This is derived from the claim in the 1950s by US economist Simon Kuznets that inequality rises in the first stages of development but then begins to fall after a plateau is reached. Controversial for a number of decades, accumulating evidence since the 1980s has discredited it (Kirby 2010: 114–115). The environmental version, proposed in the 1990s, suggests a similar relationship between environmental quality and development, namely that environmental quality will decline in early stages of development but will improve after a certain point is reached. While attempts to correlate the level of many environmental pollutants with stages of development have shown positive evidence in the case of some of these pollutants, the evidence of these two chapters discredits the EKC in terms of any correlation between levels of development and moving to low-carbon sustainability. The cases where such a correlation is evident are the exceptions rather than the rule. As Gough wrote in relation to the EKC, 'there is general agreement that it does not apply to resource use including energy and to CO_2 and greenhouse gas emissions, which tend to rise monotonically with GDP' (Gough 2016: 33).

Latin America: Challenging Development Models

Climate change is expected to have severe impacts on Latin America's development. According to a recent World Bank survey, temperatures are projected to increase by up to 4.5°C by the end of the century. The authors conclude:

> Associated physical impacts include altered precipitation regimes, a strong increase in heat extremes, higher risks of droughts and increasing aridity. Moreover, the mean intensity of tropical cyclones, as well as the frequency of the most intense storms, is projected to increase while sea levels are expected to rise by between 0.2 and 1.1 mm depending on warming level and region. Tropical glacier volume is found to decrease substantially, with almost complete deglaciation under high warming levels. (Reyer et al. 2015: 1)

As population is expected to increase from 588 million in 2013 to 660 million by 2025, reduced agricultural yields, livestock and fisheries, and species range shifts that threaten territorial biodiversity, will challenge human livelihoods as income from fisheries, agriculture and tourism decreases. These are compounded by expected negative impacts on human health, coastal infrastructures and energy systems. As a result, Latin America and the Caribbean 'will be severely affected by climate change, even under lower levels of warming, due to the potential for impacts to occur simultaneously and compound one another' (ibid.).

Though it accounts for only 9% of global carbon emissions, the social and economic impact 'is significantly greater than in the developed and emergent countries that are responsible for the bulk of those emissions because of the region's socio-territorial vulnerabilities and weak political institutions' (Spikin and Hernández 2016: 7). Yet countries in the region have been taking measures both to reduce emissions and to strengthen resilience to the impacts of climate change. Bolivia took a lead by enacting in 2010 the first law recognising the rights of nature and stating the government should develop policies to safeguard the Earth from the causes of global climate change. It has also pledged to increase the share of renewables in its energy mix from 39% in 2010 to 79% by 2030. Colombia has won awards for its sustainable transport systems in Medellín and Bogotá and is implementing a National Adaptation Plan for Climate Change. It has pledged a 20% cut in emissions by 2030. Ecuador is stemming deforestation and promoting ecosystem recovery while planning to generate 93% of its electricity from hydropower. It pledges to reduce emissions from the energy sector by between 20.7% and 25% by 2025. Guatemala has introduced a law to combat deforestation and pledges an 11.2% reduction in emissions by 2030. Peru is committed to adaptation measures to better prepare for climate disruption, focusing on water, agriculture, fisheries, forestry and health. It has pledged to reduce emissions by 20% by 2030 (WWF 2016).

Box 7.2: 'Buen Vivir': A New Development Model?

Out of the popular mobilisation with the strong involvement of indigenous peoples that has characterised politics, particularly in Ecuador and Bolivia over recent decades, has emerged 'new proposals of profound change that offer pathways to a transformation of today's civilisation' (Acosta 2013: 15). Known as Buen Vivir in Spanish (Good Living), this new approach is also known by its name in the two major indigenous languages of the region, Sumak Kawsay in Kichwa and Suma Qamaña in Aymara, and has been adopted in the Constitutions of Ecuador and Bolivia as the ultimate goal of development. As economist Alberto Acosta, who was president of Ecuador's constituent assembly drawing up a new Constitution, explains, it 'introduces an important qualitative step to overcome the traditional concept of development, much richer and more complete in content' (ibid.: 15).

It begins by recuperating the cosmovision of the indigenous peoples from which emerge some of its fundamental elements:

- Rethinking the state as plurinational and intercultural;
- Constructing a new institutionality which requires putting the citizen at the centre of the state, through communitarian spaces and active forms of social organisation;
- Overcoming the colonial and patriarchal origins of the state as well as the profound racism of society;
- Developing a social and communitarian economy through relations of production, exchange and cooperation that can provide sufficiency and quality and not just efficiency;
- Defending the right of free time for workers against forms of organisation that are ending up causing the destruction of the planet;
- Getting rid of the divorce between nature and the human which is putting at risk the very existence of humanity on Earth.

'Today, more than ever', writes Acosta, 'it is indispensable that we construct new forms of living that are not subject to the needs of capital accumulation. Buen Vivir is a contribution to this, including for its politically transformative and mobilising ability' (Acosta 2013: 19). However, as Bretón Solo de Zaldívar writes, the concept is already being appropriated by intellectuals close to the Ecuadorian government to legitimise an extractivist model of development and also by indigenous emphasising an essentialist identity (Breton Solo de Zaldívar 2016).[2]

However, these policies 'are often in conflict with prevailing economic policies and practices such as the overexploitation of natural resources, mining extractivism, deforestation, monocropping, dependence on fossil energy, rapid and unregulated urbanization and the lack of public participation in policy making and decision making about private investments that affect the environment' (Spikin and Hernández 2016: 8). This draws attention, therefore, to the region's continuing dependence for economic growth on the export both of natural resources such as oil, gas, iron ore, copper, gold, nickel, zinc, bauxite and silver, and also of foodstuffs such as sugar, coffee, soya, maize, wheat and beef. Furthermore, since the 'new left' turn in the region in the 2000s with left-wing governments taking power in most of South America and some countries of Central America, income from these exports has been relied on to fund generous social programmes such as the Bolsa Familia in Brazil that strengthened the left's base of support (Hogenboom 2012). Curbing extractivism and the export of unprocessed natural resources in order to address climate change and its impacts threatens therefore the region's dominant development model. Going even further, de la Cuadra sees this clash as an illustration of the deeper problem facing humanity since our dominant understanding and practice of development 'is founded upon an ontology and epistemology which became the civilization model for the entire human race'. Its values 'are based on a particular construction of modernity which values Western scientific knowledge above all other types of knowledge and understands progress as an essential mark based on the idea of growth and the exploitation of human and natural resources' (De la Cuadra 2015: 30).

The re-emergence of indigenous peoples in Latin America over recent decades as a cultural, social and political force has begun to challenge these dominant conceptions of development and propose alternatives based on a new balance between society and nature (see Box 7.2 on Buen Vivir). The real tensions between these two opposing conceptions of development have led to determined resistance by indigenous communities to the exploitation of the natural resources of the region, what Merchand Rojas calls 'an explosion of socio-environmental conflicts' (Merchand Rojas 2016: 172). These conflicts, which involve local communities, often indigenous, against national and transnational companies and state agents, arise over mega mining and hydroelectric projects, contamination of soil and waters, deforestation, access to and use of protected lands, dispossession of peasant communities, incursions on indigenous lands and exhaustion or contamination of fisheries. As reported by Martinez-Alier et al. (2016: 207), the number of such conflicts, and the involvement of indigenous peoples in them, is far higher in South America than anywhere else in the world. Merchand Rojas gives details of 197 environmental conflicts in 21 different countries though he doesn't state over what period of time. These include conflicts over energy resources (coal, oil and gas), mining, water, forestry and biodiversity, and agribusiness (ibid.: 175–178). The Latin American Observatory of Environmental Conflicts (OLCA.cl) keeps a comprehensive database. Two that have been identified as emblematic are the TIPNIS case in Bolivia and Camisea in Peru because they occurred in countries ruled at the time by 'new left' presidents and which claim to have policies protecting indigenous peoples (De la Cuadra 2015: 30). The TIPNIS case related to a road built to facilitate the activities of logging and oil groups through the Isiboro Sécure indigenous national park (TIPNIS), home to four indigenous communities. This lead to determined opposition by the local communities through marches, occupations, road blocks and vigils; at one point local groups kidnapped the Foreign Minister who had come to mediate. Finally President Morales suspended the project, saying he was going to consult the local communities. The second case relates to the exploitation of natural gas underground deposits in the KNN indigenous reserve which had been established in 1990 to protect the local indigenous communities. When the multinational company Shell found natural gas deposits in the region, the government facilitated them to develop infrastructure to exploit it, despite opposition from the local communities who fear serious damage to their lands and way of life. The government, however, justifies the project on the grounds of growth, job creation and benefits to the national economy.

In global climate negotiations, 'Latin American governments rarely speak with one voice on climate change or adopt common positions', state Edwards and Roberts (2015: 2). While all of them, except for Mexico, are part of the large group of developing countries known as the Group of 77+China (G77) (though it consists of 134 countries), countries from the region participate in various groups, including the Alliance of Small Island States (AOSIS; see below). Two groups have emerged from the region, ALBA and AILAC. The first, the Bolivarian Alliance for the Peoples of our America, includes Venezuela, Bolivia and Ecuador, and makes demands on developing countries to honour their environmental debt and contribute considerable sums to developing countries which have not caused the problem but are suffering its impacts. ALBA has recently joined the Like-Minded Developing Countries Group (LMDC) with China, India and various Arab states which calls on developed countries to drastically reduce their emissions and live up to pledges on climate finance and technology transfer. The Independent Association of Latin America and the Caribbean (AILAC), formed in 2012 by Colombia, Chile, Costa Rica, Peru, Panama and Guatemala, also calls on developed countries to be more ambitious but differs from ALBA in its emphasis on bold domestic action at a national level and its willingness to accept the model of carbon trading which ALBA rejects. Edwards and Roberts identify a number of competing forces that fragment Latin American responses on climate change, not least the slowdown of economic growth due to the decline of Chinese demand for the region's raw materials and leaders' fears that action on climate change 'will cost jobs and put a brake on growth' (ibid.: 3).

Africa: Adapting to Changing Conditions

Africa, with 14% of the world's population, accounts for only 0.5 tonnes of carbon emissions per capita, one twentieth that of the UK, and 2.3% of global fossil fuel consumption. Sub-Saharan Africa accounts for just 3.6% of world GHG emissions, reflecting low levels of income and energy consumption. 'The world's poorest countries are the least responsible and most vulnerable to the negative impacts of climate change', writes Michael Keating. 'With this awareness comes mounting frustration' (Keating 2009: 11). Not only is Africa's climate likely to be affected more severely by climate change than that in other regions, but its economy

Table 7.1 Selected development indicators: Latin America, Africa, Asia/Pacific

Country/Region	*GHG tonnes per capita 2013*[a]	*GDP per capita $ 2014*[b]	*Income Inequality by Gini coefficient 2010–2014*[c]	*Life satisfaction 2012–2014*[d]	*EF*[e] *per capita global hectares 2012*[f]	*How many Earths?*[g]
Latin America	–	–	–	–	2.8	1.6
Uruguay	9.98	20,886	41.6	6.7	2.9	1.7
Costa Rica	2.93	15,161	48.5	8.5	2.8	1.6
Barbados	12.55	16,114	n/a	n/a	4.5	2.6
Africa	–	–	–	–	1.4	0.8
Ethiopia	1.30	1,501	33.17	4.2	1.0	0.6
South Africa	9.59	13,128	63.38	6.3	3.3	1.9
Asia/Pacific	–	–	–	–	2.3	1.3
Bangladesh	1.04	3,132	32.13	5.3	0.7	0.4
South Korea	13.41	33,632	n/a	5.8	5.7	3.3

[a]WRI CAIT data explorer excluding LULUCF
[b]World Bank database, in purchasing power parities at current prices
[c]World Bank database
[d]worlddatabaseofhappiness.eur.nl
[e]See footnote 2, page 167.
[f]footprintnetwork.org
[g]footprintnetwork.org

is far more vulnerable to climatic variation since agriculture accounts for more than 60% of employment and, in some countries, for more than 50% of GDP. The region's economies have not shown a great ability to adapt to changes and technical progress has been slow. GHG emissions have remained low and future projections suggest that they will remain a trivial proportion of global emissions. As Table 7.1 shows, Africa lives well within its ecological footprint, the only region in the world that does so. As Collier et al. put it:

> Hence, whereas in other regions the key issues concern how to reduce carbon emissions, in Africa they concern the adaptation of production to changing, and mostly deteriorating, opportunities. Further, whereas for other regions the main adverse consequences of global warming occur only far into the future and are uncertain, in Africa many of the adverse consequence are already apparent. (Collier et al. 2008: 337–338)

Yet, as Death points out, African states 'tend to do rather badly on most indices of environmental performance and governance' since they score poorly on quality of life indicators and do not need to prioritise reducing GHG emissions since they are already very low (Death 2016: 117). This may reinforce assumptions about the weakness of African states and their lack of capacity to address environmental issues. He shifts the focus from reducing emissions to point out that many African states have long histories of managing more sustainable resource use, mitigating environmental threats and protecting sites of natural heritage. 'These include programmes of agricultural reform and tree planting, urban planning and resettlement schemes, irrigation projects and dams, massive conservation projects, disease eradication and public health programmes' (ibid.: 120). Africa hosts a number of major environmental projects, including the Grand Inga dam in the DRC, scheduled to become the world's largest source of hydropower, Morocco's plans to develop one of the largest concentrations of solar plants in the world, the construction of the Nzema solar photovoltaic plant in Ghana that will be Africa's largest and the fourth largest in the world, the Lake Turkana Wind Power Project that is one of the largest investments in wind energy in Africa and Algeria's plans to invest $60 billion in renewable power by 2030. He points to Ethiopia's strategy for a Climate Resilient Green Economy (see Box 7.3) and prominent green economy strategies that have been developed in Rwanda, Mozambique and South Africa. Instead of seeing Africa as being weak on environmental governance, he suggests that 'the African state is actually a product of particular attempts to govern land, species, human population, water resources and so on' so that what he calls 'the green African state' is not of recent origin but rather has emerged from 'long-standing and deep-rooted attempts to govern environmental resources' (ibid.: 123).

Africa has also developed its capacity to become a more influential actor in global environmental politics. While the African Group of Negotiators (AGN) was established in the early 1990s to co-ordinate African countries' positions in global negotiations on climate change and other environmental issues, up to the mid-2000s it remained weak and had little success. This was evidenced by its failure to get its concerns addressed in the design of the Clean Development Mechanism (CDM) and the subsequent failure of Africa to benefit from it. According to Roger and Belliethathan, this derived from a number of constraints, among them inadequate resources, limited access to high-quality information and poor negotiating skills (Roger and Belliethathan 2016: 93). However, the global climate summit of 2006 which took place in Nairobi marked a turning point for the AGN, resulting in it becoming a much more effective negotiating group. The summit helped to raise the profile of environmental issues among the region's leaders who channelled more resources to the talks. As a result, African negotiators were able to develop a common position that more adequately reflected the region's urgent needs. In subsequent global summits, such as at Copenhagen in 2009 and Durban in 2011, the AGN was more assertive and more influential. At Copenhagen, the efforts of the late Prime Minister of Ethiopia, Meles Zenawi, who represented Africa in the Heads of State meetings, proved decisive in the decisions made to call for $30bn in new and additional climate funding for 2010–2012 and $100bn annually by 2020. At Durban, the AGN played 'a crucial role' in supporting the extension of the Kyoto Protocol and the decision to negotiate a new global climate treaty to be binding on all states and to come into force in 2020 (ibid.: 101). This Durban Platform for Enhanced Action as it is called formed the basis for the negotiation of the 2015 Paris Agreement.

However, despite the fact that the average African has an EF of only 1.4 global hectares (gH), the lowest of the major world regions, some countries do have significant emissions and need to reduce them. South Africa, with an EF of 3.3gH is one of the highest in Africa (only Botswana (3.8gH), Libya (3.7gH) and Mauritius (3.5gH) are higher) but its population at over 52 million dwarfs those of the other high emitters. Therefore, it is interesting to examine South Africa's reduction pledges. In its INDC, it pledged to reduce its emissions to between 398 and 614 million tonnes of CO_2 equivalent over the period 2025 to 2030. However, Climate Action Tracker estimates this is equivalent to an increase of between 20% and 82% on its 1990 levels (climateactiontracker.org/countries/southafrica) and it rates it as 'inadequate' to reach the global target of keeping warming within 2°C. Indeed, if other countries did the same, global warming would reach between 3°C and 4°C, it states. The main problem for South Africa is its reliance on coal to generate the electricity required by industry and construction. While it is developing renewable energy projects, the Climate Action Tracker states that this will not displace coal generation which is growing at a similar rate to renewable capacity.

Box 7.3: Ethiopia 'An Exemplary Case of Low-Carbon Development'
Ethiopia, with a population of 92 million and an average per capita income of $574 a year, is being identified as an 'exemplary case for other low-income countries' as it is one of the few countries promoting low-carbon development (Urban et al. 2013: 228). Though not a notable contributor to climate change since its people's average EF is just 1gH, it is very vulnerable to its effects being largely dependent on agriculture, forestry and fisheries. Already climate change has resulted in periodic droughts and flooding, damaging food production and water supply.

The government is now committed to the twin aims of becoming a middle-income country and carbon neutral by 2025. Through its ambitious energy policy to develop up to 21 hydropower projects and large-scale wind and solar power capacity, it hopes to supply electricity to the 85% of urban dwellers and the 95% of rural dwellers who do not currently have it. It also wishes to promote fuels with a lower-carbon content by developing gasohol, a blend of gasoline and ethanol. Through its extensive reforestation programme and improved land-use and livestock management, it hopes to develop a low-emissions agriculture. Government and state-owned enterprises are seen as the main drivers of the policy though they are encouraged by international donors.

However, a number of possible trade-offs are feared. The first is the impact on agriculture which is the main source of economic growth. Might a move to carbon neutrality result in reduced growth? Secondly, it is feared that significant investment in renewables might come at the expense of investment in poverty-reduction. Thirdly, will the development of biofuels be at the expense of food production? Fourthly, will a greater reliance on hydropower make energy generation more vulnerable? Furthermore, the whole programme remains dependent on international financial aid and technological transfer, as does Ethiopia's INDC pledge to reduce emissions by at least 64% by 2030. CAT assesses Ethiopia's targets as 'sufficient' but if they were not conditional it would rate it a 'role model' (climateactiontracker.org/countries/ethiopia).

Asia: Development Versus Sustainability?

Asia is the region of the world where climate change is expected to have some of the most devastating impacts. The Asian Development Bank draws attention to the fact that many of its major cities are low-lying or coastal and therefore highly vulnerable to rising sea levels, floods and other impacts of climate change. These include Bangkok, Dhaka, Guangzhou, Ho Chi Minh City, Kolkata, Manila, Mumbai, Shanghai and Yangon. Furthermore, consuming 80% of the region's energy and creating 75% of GHG emissions, Asian cities will contribute more than half the rise in global GHGs in the next 20 years if no action is taken (ADB 2016) (see Box 7.4 on Asian cities building resilience). The World Bank highlights that parts of South East Asia are located within a tropical cyclone belt and are characterised by archipelagic landscapes and relatively high coastal population density. This makes the region particularly vulnerable to sea-level rise, increases in heat extremes, increased intensity of tropical cyclones and ocean warming and acidification. For example, with warming levels of 1.5°C to 2°C it predicts that sea-level rise will reduce rice production in the Mekong River Delta in 2040 by about 2.6 million tons a year or about 11% of 2011 production while population will increase from 590 million in 2010 to around 760 million by 2050. Marine fishing stocks are projected to halve off the southern Philippines by 2050 due to warmer sea temperatures and ocean acidification (World Bank 2013: 65). In South Asia, comprising the populous countries of India, Pakistan and Bangladesh, projected changes in temperatures, rainfall and the frequency and intensity of weather events will impact on monsoon activity, droughts, glacial loss, snow levels, river flow, ground water resources and sea-level rises. 'An abrupt change in the monsoon ... toward a drier, lower rainfall state, could precipitate a major crisis in South Asia' (ibid.: 108) while the retreat of the Himalayan glaciers could seriously affect food production for 750 million people. Total crop production is expected to decrease with calorie availability declining resulting in significant health problems. Childhood stunting is projected to increase by 35% by 2050 due to climate change. The World Bank concludes: 'The region is highly vulnerable even at warming of less than 2°C given the significant areas affected by droughts and flooding at present temperatures. In addition, the projected risks of crop yields and water resources, and sea-level rise reaching 70cm by the 2070s, are likely to affect large populations' (ibid.: 109).

Yet Asia is also the region in which some countries are seen as forerunners in decoupling emissions from growth, thereby leading the way in green growth and holding lessons for the rest of the world. The Asian Development

Bank has singled out South Korea, Japan and Singapore in this regard. For the UNEP, South Korea is 'a model green-growth nation' due to the pivotal role played by the Lee Myung-bak administration (2008–2013) in propagating green growth with 'the potential of starting a domino effect on the major Asian economies' (quoted in Bluemling and Sun-Jin-Yun 2016: 115–116). Indeed, the term 'green growth' first emerged in intergovernmental discussions in the South Korean capital Seoul at the Fifth Ministerial Conference on Environment and Development in Asia and the Pacific in 2005 leading to the establishment of the Seoul Initiative Network on Green Growth. The UN Economic and Social Commission for Asia and the Pacific (UNESCAP) defined it in 2006: 'Green growth proposes to harness the power of economic growth while guiding it in a way that will enhance the immense possibilities provided by innovative technologies and industries, so that progress can be registered in more than gross domestic product increases alone' (quoted in ibid.: 118). It was taken up as its national development strategy by the incoming Lee Myung-bak government in 2008 and became central for domestic policy combining ecosystem protection (such as major dam projects) with economic growth. It also involved a major expansion of nuclear power capacity. The administration of Park Geun-hye which took power in 2013 (President Park was impeached and left power in 2017) devised a new version of the strategy, Green Growth 2.0 with the emphasis on the 'creative economy' and a more bottom-up approach. GHG emissions in Seoul are to be reduced by 25% by 2020 and 40% by 2030. Critics, however, point to the technocratic and capital-intensive nature of the green growth strategy in South Korea and question its claims to have decoupled growth from emissions, pointing to the fact that energy consumption is increasing faster than economic growth (ibid.: 125).

Neither do the INDCs of South Korea, Japan and Singapore attest to them being leaders in reducing emissions, as analysed by CAT. It finds South Korea's pledge to reduce emissions by 37% below business-as-usual emissions by 2030 to be inadequate as it is not in line with keeping global warming to below 2°C. It points out that South Korea's emissions have more than doubled between 1990 and 2012 and it is one of the fastest growing emitters. 'The high export rates from Korea's manufacturing industry play a critical role in Korea's increasing emission levels', says the CAT, and it is unlikely to meet its 2020 pledge with current policies (climateactiontracker.org/countries/southkorea). Japan pledged to reduce emissions by 26% below 2013 levels by 2030 though this is in part relying on the use of carbon credits. CAT assesses this as a serious decrease in ambition compared to its pledge at the Copenhagen summit in 2009. If all countries followed Japan's example, it would lead to global warming in excess of 3–4°C, says CAT. With coal-fired power plants set to play an increasingly important role in its energy sector, the share of low-carbon options will increase only slightly. Furthermore, its programme of installing efficient coal power stations in developing countries 'could degrade global efforts to decarbonise the energy system' (carbonactiontracker.org/countries/japan). Singapore pledged to reduce its total GHG emissions per unit of GDP by 36% below 2005 levels by 2030 with the aim of peaking its total emissions around the same time. However, this target would result in an increase in emissions of 39% above 2010 levels by 2030, according to CAT and is inadequate to keep global warming below 2°C. It adds that 'Singapore is one of the world's largest international navigation and aviation hubs. Emissions associated with these activities are nearly three times as high as domestic emissions and have been rising steeply over recent decades' (carbonactiontracker.org/countries/singapore). The country has replaced oil with natural gas in its power sector but places little emphasis on renewables and while the emissions from natural gas are lower than those from oil, 'the gas liquefaction, transportation and regasification significantly increases the carbon footprint of gas', says CAT. Furthermore, as one of the largest oil refining and oil products trading hubs in the world, the growth of this sector over coming decades will further increase the country's emissions.

Box 7.4: Asian Cities Build Resilience

Recognising that the impacts of climate change are unique to local conditions and that the ability to address them depends on local capacities, governance structures and resources, cities around the world have come to play a leading role in planning for sustainability. For example, the C40 cities network has 86 cities affiliated to it in all regions of the world, comprising 25% of global GDP and containing one in 12 of the world's population (c40.org). In Asia, the Rockefeller Foundation has funded the Asian Cities Climate Change Resilience Network (ACCCRN), including 10 cities in Vietnam, Indonesia, India and Thailand undergoing rapid growth.

Begun in 2008, this network sees cities as 'dynamic systems where building urban resilience is a process of evolutionary transformation'

(Kernaghan and da Silva 2014: 48). The project, which ran to 2016, involved four phases:

- Knowledge: raise awareness of the challenges and develop networks of learning across the ten cities;
- Stakeholders: draw in a wider group of stakeholders including city champions and entrepreneurs, government leaders, academia, the private sector and civil society;
- Policies and Plans: develop city-wide plans including multiple sectors such as land use planning, energy management, ecosystem services, housing and transport, water supply and sanitation, health services, education and waste management, embedding climate change within an overarching strategy;
- Finance: looking for alternative sources of funds to sustain action into the future.

While clear progress has been made in mobilising a variety of stakeholders and in developing actions to improve urban resilience, major challenges remain. The efficacy of the actions undertaken is not yet proven and it has been difficult to draw in the private sector which remains focused on their own businesses or property. Getting commitment from government leaders at all levels from the local to the national is also challenging. But the most critical issue remains finance, both accessing sources of funding and doing so on a scale needed to address the problem (Kernaghan and da Silva 2014).

It must be concluded therefore that, while parts of Asia face some of the most devastating impacts of climate change on highly vulnerable populations, the region's most developed states are failing to meet their obligations to fashion low-carbon development pathways, despite a rhetorical commitment to green growth strategies.

Small Island Developing States: Vulnerability and Responsibility

A group of SIDS are some of the most vulnerable countries in the world to the impacts of climate change. Made up of 52 territories across the Atlantic (Africa and Caribbean), Indian and Pacific oceans, the particular vulnerabilities of SIDS were recognised since the 1992 Earth summit: small size constraining economic development and diversification, isolation increasing costs and hindering participation in global supply networks, and exposure to disasters such as sea-level rise, tropical cyclones and hurricanes. Although contributing only 0.05% of global GHG emissions, the SIDS are among the top 'hot spots' of the world measured by frequency and severity of disasters. The impact of natural disasters on islands' infrastructures results in levels of destruction much more severe than in other regions of the world: for example, Hurricane Ivan in 2004 damaged 90% of the housing stock on Grenada with an estimated cost of US$527million or 38% of the country's GDP. Inundation of sea water also affects drinking water and agricultural production, and temperature rises and increased ocean acidification destroy coral reefs and fishery habitats (Alfaro-Pelico 2012: 5). The UN estimates that without meaningful adaptation and mitigation measures, the Caribbean region could lose 2% to 3% of its GDP annually due to climate change impacts (UNECLAC 2010). The greatest threat, however, arises from the trends in sea-level rises which pose risks to the very existence of these states. While global average sea-level rises of 1.3–2.3 mm per year were forecast by the IPCC, some SIDS are experiencing over double this level; for example, the Maldives and the south of Trinidad are experiencing a 4 mm rise annually (ibid.: 4).

To represent the unique interests of these small states in international climate negotiations, the AOSIS was established in 1990 and currently has 44 member and observer states spread over the Atlantic, Pacific and Indian Oceans. They constitute 28% of developing countries, 20% of the UN's membership but only 5% of the global population. It acts as an ad hoc lobby and negotiating voice for SIDS within the UN system (aosis.org). Though such a group of very small states has little structural power in the global system, it has very effectively managed to advance its agenda and achieved some notable successes in international climate change negotiations. Due to the vulnerability of its members to rising sea levels, it has forcefully promoted the goal of maintaining global warming to 1.5°C rather than the 2°C which has tended to dominate international discussions. As the Maldives representative said, speaking on behalf of the AOSIS at the end of the Paris climate summit, 'limiting the global rise in temperature to 1.5 degrees is a life or death matter for our most vulnerable members' (AOSIS 2015). The mention in the Paris Agreement of the commitment 'to pursue efforts to limit the temperature increase to 1.5°C above pre-industrial levels' is in part due to the lobbying efforts of AOSIS. Even more so, the recognition in Article 8 of the Paris Agreement of 'the importance of averting, minimizing and addressing loss and damage associated with the adverse effects of climate change, including extreme weather events and slow onset events' owes much to the efforts of AOSIS. What is surprising is the ability of a group of such small and effectively powerless states to get significant commitments from the global community. How did they achieve them?

In analysing the leadership of AOSIS, de Águeda Corneloup and Mol argue that these states 'turned morality into their main asset and core leadership strategy': 'Morality is used to build (discourse) coalitions and support, and to delegitimize and shame other states with different discourses, positions, and interests' (De Águeda Corneloup and Mol 2014: 292). They identify the means used to exercise such a moral leadership. First is the building of a clear discourse and a wide coalition to support that discourse. Second is the use of various 'leadership strategies'. Entrepreneurial strategies included raising SIDS' visibility and influence through hosting meetings, participating in preparatory negotiations, organising campaigns and activities, building coalitions and linking with scientists and NGOs. Intellectual strategies helped build credibility for their case through deploying scientific evidence that elevated it beyond the merely political.

> With hardly any contribution to global warming, SIDS are victims of powerful developed and emerging economies, which make current lax climate change measures morally unjust and in need of fundamental adjustment, toward a 1.5°C target, additional adaptation funds for especially SIDS, and a legally binding document. This moral 'leadership' strategy provided SIDS with most leverage in the negotiations, also because it managed to attract a broad coalition. (ibid.: 294)

Box 7.5: The Maldives Faces Conflicting Options

SIDS were the first to ratify the Paris Agreement. Among them was Maldives, a small island state of some 90,000 square kilometres in the Indian Ocean south of India with a population of around 400,000. It has pledged unconditionally to reduce its GHG emissions by 10% before 2030, a retreat from the commitment made in 2009 to become carbon neutral by 2020, under the previous administration of Mohamed Nasheed (2008–2012). However, if enabled by financial assistance, technology transfer and capacity building, it could increase its emissions reductions to 24%. It would do this by moving to renewable sources of energy.

The country's energy demand is completely met by imported fossil fuels on which it spends 30% of its GDP. Despite this, its emissions constituted only 0.003% of global emissions in 2011. But the Maldives INDC makes clear the problems posed in moving to renewables: 'Solar irradiance is available in the country throughout the year, however lack of technical capacity, limited land area, already established diesel-based power generation systems and high investment costs pose a major challenge to the introduction of solar PV systems in the country' (Naish 2015).

As well as measures to mitigate emissions, the INDC outlines adaptation measures to address the impacts of climate change such as coastal protection, enhancing food and water security, improving the resilience of critical infrastructure, and safeguarding coral reefs. Meanwhile, with the confirmation of oil and gas finds in its territorial waters, the government is facing criticism over its plans for oil exploration. This is seen by local NGOs as hypocritical as the Maldives is one of the countries most vulnerable to climate change.

As the example of the Maldives shows (Box 7.5), the challenge for many SIDS is more in the area of adaptation than in mitigation. Already emitting tiny amounts of GHG emissions and facing major challenges in moving to renewables, the focus of action for many countries is on measures to limit the impacts of climate change. The gravest risks facing small Caribbean island states is the potential damage to their tourism industry, one of the mainstays of their economies. Rising sea levels pose a particular risk due both to their long coastlines relative to their land area, and to the fact that large proportions of the land are low-lying. With a sea level rise of half to one metre, many countries would lose a significant part of their land where most tourist attractions are sited. The Caribbean heads of government in 2005 established the Caribbean Community Climate Change Centre which has been developing adaptation plans to protect key tourism assets and the coastal ecosystem on which its fishing industries depend. These include coastal protections and siting buildings further back from the coast, restoring beaches and coral reefs, sewage treatment and water management policies. Yet, as Mycoo reports, survey evidence from Barbados reveals that hoteliers are more concerned with economic survival by keeping occupancy levels high rather than the threats posed by climate change. Also included is capacity building and institutional reform to ensure databases to aid decision making, monitoring and enforcement. Much work remains to be done but the fundamental question is who will pay for it (Mycoo 2014).

Conclusions: Lessons for a Political Economy of Low-Carbon Development

The focus of discussion in these two chapters has been on pathways to a low-carbon society rather than the more usual focus on mitigation and adaptation in discussions of transition, and on the immense threats to development and wellbeing faced by the world's poorest if the transition is not successfully achieved. The benefit of focusing on pathways is that it allows mitigation and adaptation be considered as part of wider development strategies, thus bringing issues such as consumption, economic development and governance centrally into the discussion rather than the insufficient attention they often get, as has been shown in Chap. 5. The conclusions of Chap. 6 have already highlighted how market-led approaches have proven largely unsuccessful in combining real sustainability with development in most of the developed world and in key emerging economies, except where state intervention has ensured that the interests of citizens are not subordinated to those of capital as in the Nordic states. This chapter, in which the combination of sustainability and development has been a more central focus, reinforces this conclusion but provides material to examine more closely many different political economy interrelationships and draw lessons from them.

Two small countries, Costa Rica and Uruguay, emerged as models early in the chapter. Both illustrate how the state's role has successfully fostered pathways towards low-carbon development through predictability in policy and positive partnerships between the public and the private sectors. Ethiopia, at a very different stage of development and in a very different context, offers similar lessons. A major conclusion to be drawn from these cases is that the legacies of history, what is called path dependency, matters as a directive state has been developed over decades in each of these countries. The long history of African states in managing more sustainable resource use mentioned in this chapter is another example of path dependency, though with more limited influence on state capacity. These examples offer more evidence to support conclusions drawn by Sommerer and Lim: 'Environmental pioneering implies both capacity and strong political commitment. Therefore, the existence of non-Western pioneers suggests that the evolution of an environmental state ... is more than just the emulation of a Western model' (Sommerer and Lim 2016: 107). Global leadership given by the AOSIS, as outlined in this chapter, offers another example of political commitment though with extremely limited capacity. While issues of capacity and political commitment are important to highlight, the cases identified in this chapter draw attention to an equally important lesson to be drawn that can often be lost when the focus is too narrowly and exclusively on the state. This relates to the role of civil society activism in shaping and reshaping states: each of the three cases mentioned here relate to states in which strong civil society activism, at times marking revolutionary breaks with the past, have decisively shaped the national state. In each case, socialist influences in different ways, played a role.

The role of civil society has also emerged in this chapter as a significant player, especially in the case of Latin America. Here it is important to look beyond the many environmental conflicts and the ways in which they are influencing state actions and policies, to examine the different models of development that are in dispute—the dominant, capital-intensive, resource extractive and export-oriented model versus the alternative, community-led, participatory, locally based and environmentally balanced model that is emerging and being formulated in approaches such as Buen Vivir (Box 7.2). The chapter has also shown how the dominant development model in the more developed countries of Asia like South Korea, Japan and Singapore is clashing with and undermining attempts to transition to low-carbon development. Even in a small island state like the Maldives, this is emerging as a crucial issue (Box 7.5). This concurs with a point made by Duit et al. that 'much remains to be done to appreciate the social forces influencing the strength and shape of ... transitions, as well as the form of the environmental state and its interactions with society in different contexts' (Duit et al. 2016: 15–16). For, perhaps often obscured behind discussion of development models, lie very powerful social interests, namely those of business. This brings the role of the market into explicit focus and the influence of business interests in shaping environmental policy. In recognising this, Gough makes a distinction between 'brown' and 'green' capitalist interests, with the former opposing environmental measures (and finding very clear expression in the policies of the Trump administration in the US) but the latter also shaping state policies towards 'green growth' in a distinctive way, sometimes labelled ecological modernisation. As Gough writes: 'Green and climate-change agendas have largely risen in the era of dominant neo-liberal ideas, a denigration of state capacities and hostility to public initiatives' (Gough 2016: 34); their influence in shaping the transition has been clearly identified in this chapter in countries like South Korea and Singapore. The reluctance of business to become active participants in multistakeholder projects to promote low-carbon development is also illustrated in the case of some Asian cities (Box 7.4).

An overall conclusion that can be drawn from the survey in these two chapters is that the political economy models fashioning pathways towards a post-carbon society in most parts of the world display far too much deference towards market actors, devote inadequate attention to the development of state capacity and leadership commensurate with the immense challenges involved, and fail to develop the sorts of state-civil society partnerships that might galvanise the radical shifts needed. As Gough states, these radical shifts 'would challenge dominant interests and ideas' such as consumer sovereignty and unquestioned economic growth, would 'need to be accomplished in the face of accumulated policy legacies', and would require 'extensive consensual policymaking involving key constituencies of interest to set the frameworks for markets' (Gough 2016: 42). Bearing in mind these immense challenges, Chap. 8 examines the dominant political economy model through which the transition is currently being attempted, namely climate capitalism. Following this, Chap. 9 identifies an emerging alternative model which it labels ecosocialism.

Notes

1. Defining the terms 'development' and 'sustainability' generates major debates. In this chapter the following working definitions are adopted: development: 'actions to improve the living conditions of the poorest throughout the world through economic and social interventions at macro and micro levels'; sustainability: 'forms of human economic activity and culture that do not lead to environmental degradation' (Robertson 2014: 5). For discussion of the meanings of development, see McGillivray 2008 and Nederveen Pieterse 2001. For discussion of sustainability, see Robertson 2014.
2. Translation from Spanish by P. Kirby.

References

Acosta, Alberto. 2013. *El Buen Vivir: Sumak Kawsay, una oportunidad para imaginar otros mundos*. Barcelona: Icaria.

Alfaro-Pelico, Raúl I. 2012. Small Island Developing States and Climate Change: Effects, Responses and Positions Beyond Durban (WP). *Working Paper 1/2012*, Real Instituto Elcano, Madrid.

AOSIS. 2015. Statement Delivered by the Maldives on Behalf of the Alliance of Small Island States (AOSIS) at the Closing Subsidiary Body for Scientific and Technical Advice (SBSTA) Plenary, Paris, 2015. Accessed 23 September 2016. http://aosis.org/wp-content/uploads/2015/12/Paris-Closing-SBSTA-1546.pdf

Asian Development Bank. 2016. Asian's Booming Cities Most at Risk from Climate Change. Accessed 12 October 2016. www.adb.org/news/features/asias-booming-cities-most-risk-climate-change

Bluemling, Bettina, and Sun-Jin Yun. 2016. Giving Green Teeth to the Tiger? A Critique of "Green Growth" in South Korea. In *Green Growth: Ideology, Political Economy and the Alternatives*, ed. Gareth Dale, Manu V. Mathai, and Jose A. Puppim de Oliveira, 114–130. London: Zed Books.

Breton Solo de Zaldívar, Victor. 2016. Buen Vivir (Sumak Kawsay), ¿alternativa al desarrollo occidental? *e-dhc* 6: 28–41.

Collier, Paul, Gordon Conway, and Tony Venables. 2008. Climate Change and Africa. *Oxford Review of Economic Policy* 24 (2): 337–353.

De Águeda Corneloup, Inés, and Arthur P.J. Mol. 2014. Small Island Developing States and International Climate Change Negotiations; The Power of Moral "Leadership". *International Environmental Agreements* 14: 281–297.

De la Cuadra, Fernando. 2015. Indigenous People, Socio-Environmental Conflict and Post-Development in Latin America. *Ambiente & Sociedade* XVIII (2): 23–40.

Death, Carl. 2016. Green States in Africa: Beyond the Usual Suspects. *Environmental Politics* 25 (1): 116–135.

Duit, Andreas, Peter H. Feindt, and James Meadowcroft. 2016. Green Leviathan: The Rise of the Environmental State? *Environmental Politics* 25 (1): 1–23.

Edwards, Guy, and Timmons Roberts. 2015. Latin America and UN Climate Talks: Not in Harmony. *Americas Quarterly*, Winter 2015. Accessed 9 October 2016. http://www.americasquarterly.org/content/latin-america-and-un-climate-talks-not-harmony

Gough, Ian. 2016. Welfare States and Environmental States: A Comparative Analysis. *Environmental Politics* 25 (1): 24–47.

Hogenboom, Barbara. 2012. The Return of the State and New Extractivism: What about Civil Society? In *Civil Society and the State in Left-Led Latin America*, ed. Barry Cannon and Peadar Kirby, 111–125. London: Zed Books.

Keating, Michael. 2009. With One Voice. *The World Today*, October 2009, 10–11.
Kernaghan, Sam, and Jo da Silva. 2014. Initiating and Sustaining Action: Experiences Building Resilience to Climate Change in Asian Cities. *Urban Climate* 7: 47–63.
Kirby, Peadar. 2010. *Celtic Tiger in Collapse: Explaining the Weaknesses of the Irish Model*. Basingstoke: Palgrave Macmillan.
Lamb, William F. 2016. Which Countries Avoid Carbon-Intensive Development? *Journal of Cleaner Production* 131: 523–533.
Martinez-Alier, Joan, Federico Demaria, Leah Temper, and Mariana Walter. 2016. Trends of Social Metabolism and Environmental Conflict: A Comparison Between India and Latin America. In *Green Growth: Ideology, Political Economy and the Alternatives*, ed. Gareth Dale, Manu V. Mathai, and Jose A. Puppim de Oliveira, 187–210. London: Zed Books.
McGillivray, Mark. 2008. What is Development? In *International Development: Issues and Challenges*, ed. Damien Kingsbury, John McKay, Janet Hunt, Mark McGillivray, and Matthew Clarke, 21–50. Basingstoke: Palgrave Macmillan.
Merchand Rojas, Marco Antonio. 2016. Neoextractivismo y conflictos ambientales en América Latina. *Espiral, Estudios sobre Estado y Sociedad* XXIII (66): 155–192. May–August 2016.
Mycoo, Michelle. 2014. Sustainable Tourism, Climate Change and Sea Level Rise Adaptation Policies in Barbados. *Natural Resources Forum* 38: 47–57.
Naish, Ahmed. 2015. Maldives Pledges 10 Percent Reduction in Carbon Emissions by 2030. *Maldives Independent*, 29 September 2015.
Nederveen Pieterse, Jan. 2001. *Development Theory: Deconstructions/Reconstructions*. London: Sage.
Pashley, Alex. 2016. Uruguay Pushes 100% Renewables, Just Don't Mention the Cows. *Climate Change News*, 19 February 2016. Accessed 30 September 2016. www.climatechangenews.com/2016/02/19
Pope Francis. 2015. *Laudato Si: On Care for our Common Home*. Vatican City: Vatican Press.
Reyer, Christopher P.O., Sophie Adams, Torsten Albrecht, Florent Baarsch, Alice Boit, Nella Canales Trujillo, Matti Cartsburg, et al. 2015. Climate Change Impacts in Latin America and the Caribbean and their Implications for Development. *Regional Environmental Change*, October 2015. Accessed 7 October 2016. doi:10.1007/s10113-015-0854-6.
Robertson, Margaret. 2014. *Sustainability: Principles and Practice*. London: Earthscan from Routledge.
Roger, Charles, and Satishkumar Belliethathan. 2016. Africa in the Global Climate Change Negotiations. *International Environmental Agreements* 16: 91–108.
Sommerer, Thomas, and Sijeong Lim. 2016. The Environmental State as a Model for the World? An Analysis of Policy Repertoires in 37 Countries. *Environmental Politics* 25 (1): 92–115.

Spikin, Andrea Santelices, and Jorge Rojas Hernández. 2016. Introduction: Climate Change in Latin America: Inequality, Conflict, and Social Movements of Adaptation. *Latin American Perspectives* Vol. 43-4: 4–11.

UNECLAC. 2010. Review of the Economics of Climate Change in the Caribbean Project. Port of Spain: UN Economic Commission for Latin America and the Caribbean.

Urban, Frauke, and Johan Nordensvärd. 2013. Approaches to Low Carbon Development in Low, Middle and High Income Countries. In *Low Carbon Development: Key Issues*, ed. Frauke Urban and Johan Nordensvärd, 217–227. London: Earthscan from Routledge.

Urban, Frauke, Marie Blanche Ting, and Hilawe Lakew. 2013. Poverty Reduction and Economic Growth in a Carbon Constrained World: The Case of Ethiopia. In *Low Carbon Development: Key Issues*, ed. Frauke Urban and Johan Nordensvärd, 228–239. London: Earthscan from Routledge.

Watts, Jonathan. 2015. Uruguay Makes Dramatic Shift to Nearly 95% Electricity from Clean Energy. *The Guardian*, 3 December 2015.

World Bank. 2013. *Turn Down the Heat: Climate Extremes, Regional Impacts, and the Case for Resilience*. Washington, DC: The World Bank.

WWF. 2016. Latin America and the Caribbean Action on Climate Change. Accessed 9 October 2016. http://www.worldwildlife.org/climatico/latin-america-and-the-caribbean-take-action-on-climate-change

PART IV

Pathways to a Low-Carbon Future

CHAPTER 8

Climate Capitalism: How Far Can It Get Us?

The previous two chapters examined the pathways to a low-carbon society currently being taken by various countries in different regions of the world. The conclusions drawn highlighted the importance of political economy models to the outcomes achieved, identifying such key dimensions as the respective role of state and market and the interactions between them, the role of civil society and the fact that these models are based on the longer-run contours of the development models that emerged through power struggles over the past century. As was highlighted, it was those countries that fashioned a political economy model capable of achieving successful developmental outcomes (for e.g., Costa Rica and Uruguay among developing countries) that are proving most successful in opening pathways towards a post-carbon world that combine improvements in quality of life indicators for most of their citizens with GHG emissions reductions. The main lesson to be drawn is that success in moving towards a post-carbon society requires a development model adequate to the task.

This section therefore examines political economy development models for a post-carbon society. This chapter identifies and analyses the dominant model currently in place to achieve this transition, namely climate capitalism. The following chapter outlines key dimensions of an alternative model that finds expression in micro projects around the world and in the work of analysts. This we label an ecosocialism. The final chapter highlights the options now facing the global community and the prospects for

P. Kirby, T. O'Mahony, *The Political Economy of the Low-Carbon Transition*, International Political Economy Series,
DOI 10.1007/978-3-319-62554-6_8

which pathways will be taken. The focus in these chapters therefore is on the broader contours of the dominant model, which is climate capitalism, or a possible alternative emerging model such as ecosocialism. Within these, there can be many different emphases that affect outcomes, most especially the role of public authorities in developing robust policies for the transition as against leaving it largely to free market actors. But behind these differences lie commonalities that can restrict actions or open out broader pathways. These chapters seek to identify such commonalities that define a dominant model currently structuring the pathways being taken, and the contours of a possible alternative model that could structure those pathways in a very different, and more successful, way.

The chapter begins by defining what is climate capitalism and describing its key features. The following section takes a few examples of how it operates in practice, drawing on the reports of the New Climate Economy project, and on the concept of green growth as promoted, for example, by the OECD, the UNEP and the World Bank. Each of the following three sections examines in turn a major challenge for climate capitalism. The first of these takes the key question of finance, situating the challenges for climate action in a wider analysis of the financialisation that characterises contemporary capitalism. The second turns to governance, drawing on a detailed report of the German Advisory Council on Global Change. The third and major challenge for climate capitalism is the issue of decoupling growth from GHG emissions, a necessary condition for its success. The final section draws conclusions about how far the model of climate capitalism can take us on the pathway to a post-carbon society, raising questions about the central role played by the market and about economic growth itself.

What Is Climate Capitalism?

Newell and Paterson (2010) introduced the term 'climate capitalism' as a way of identifying how 'an embryonic form of climate capitalism is already emerging' as governments, corporations and non-governmental actors are responding to climate change (Newell and Paterson 2010: 8). Though not discussing it explicitly as a political economy model, they do define this form of capitalism as 'a model which squares capitalism's need for continual economic growth with substantial shifts away from carbon-based industrial development' (ibid.: 1). They further situate the emergence of climate capitalism within the characteristics that define neoliberalism: 'the ideological fixation with *markets*, the dominance of *finance*, the widening global economic *inequalities*, and the focus on *networks* as a means of organising' (ibid.: 23–24; emphasis in original). All have combined to shape the character of capitalist responses to climate change, they argue. The dominant form of climate capitalism is therefore a variant of neoliberal capitalism, seeking to use its mechanisms to address climate change. Studying the climate and environmental policy groups (CEPGs) that promote this form of capitalism, Sapinski makes the further point that 'the crux of climate capitalism is to make reducing greenhouse gas emissions profitable for a large enough number of corporations that would give their political support to the project' (Sapinski 2016: 90).

Reconciling economic growth with achieving environmental goals countered the discourse that emerged in the 1970s identifying environmental limits to growth (see Box 5.2). As firstly outlined in Meadows et al. (1972) and subsequently updated in Meadows et al. (1992 and 2004) and in Randers (2012), the limits to growth thesis concluded that humanity was poised to grow dangerously beyond the physical limits of the planet resulting in two options, either managed decline or collapse. As Randers put it in his 40-year update, 'overshoot and collapse were a future possibility that my colleagues and I really believed would be avoided through new, wise, and forward-looking policy'. He adds: 'Sadly, though, it is not obvious that the last forty years has given support to our youthful optimism' (Randers 2012: 3). Against the 'command and control' policies being advocated to keep humanity within planetary limits (such as regulations or prohibitions), economists promoted market mechanisms to achieve environmental goals. The two principal mechanisms are environmental taxes and emissions trading schemes (ETS); in each case, the cost of emitting carbon creates incentives for changing behaviour to avoid emissions (Smith 2011: 40–50). Decisions are therefore taken away from governments and regulators and left to individuals and companies. Fitting well with the main elements of neoliberalism, these ideas 'left a powerful impact' in the late 1980s (Newell and Paterson 2010: 25).

In keeping with the priority given to efficiency over equity within currently applied neoliberal thinking, emissions trading came to be preferred over carbon taxes. In the EU, a long debate on carbon taxes met strong industry resistance and the reluctance of some member states to cede tax-raising powers to Brussels. First proposed by the US in 1996, it was the EU that created the first and still the largest ETS following negotiations on the Kyoto Protocol in 1997. However, its subsequent development has been fraught with difficulties. The system works by setting a cap on the total amount of certain GHGs that can be emitted by those entities covered (more than 11,000 heavy energy-using power stations and industrial plants, and airlines operating between the countries covered by the system). Since the cap is reduced over time, emissions are expected to also decline. Companies receive or buy emissions allowances that they can trade with one another as needed; they can also buy a limited amount of credits from outside the system but the limit on the total amount of allowances available is designed to ensure their value is maintained. Companies that fail to live within their allowances are heavily fined; those with spare allowances can trade them or keep them for future use. The EU ETS is now in its third phase, covering the period 2013–2020, introducing a single EU-wide cap to replace earlier national caps, auctioning most allowances instead of offering them free, drawing more sectors and greenhouse gases into the system, and setting aside 300 million allowances to fund the deployment of innovative renewable energy technologies; many of these reforms are designed to boost the price of carbon within the system. The ETS operates in the 28 EU member states as well as Iceland, Liechtenstein and Norway, and covers around 45% of the EU's GHGs. It accounts for over three-quarters of international carbon trading though new systems are being established around the world (New Zealand, US, Canada, China, Australia and Japan, either at national or subnational levels). Figures released in mid-2016 showed an overall decline of 0.37% in emissions in 2015 over the previous year but an increase of 3.6% in aviation emissions in that period (EC 2016). However, the price for a ton of carbon in the system in March 2017 was hovering just above €5.

Assessment of the effectiveness of the ETS system tends to focus on a series of trade-offs: environmental outcomes at limited cost to the economy, a price signal that spurs behavioural responses and innovation, and revenue to finance mitigation and adaptation measures. Evidence seems to show that the EU ETS has been a contributor to meeting reduction targets as emissions from the sectors covered by the ETS have decreased by 24% over 2005 levels, beating the target of a 21% decrease by 2020. However, how much of this is due to the severe economic recession suffered by many EU countries reducing production over this period, how much is actually due to the operation of the ETS? Marcu et al. (2016) state that data availability does not allow an independent analysis to verify this; furthermore, they add that these targets are not sufficient to meet the objectives of the Paris Agreement (PA). The more ambitious objective set by the PA makes the current EU contribution to the global reduction effort insufficient, and it already 'has a low probability (66%) of achieving the 2°C target' (Marcu et al. 2016: 6). Laing and Mehling report evidence that, despite the low price for carbon, 'the mere existence of an incentive to reduce carbon is helping to change decision-making in some corporate entities' (Laing and Mehling 2013: 9). A major issue is the lack of incentive posed by the low carbon price. Attempts to address this by the elimination of the surplus of allowances in the system 'resulted in the outcome being a watered down compromise' that has not had the desired result (Marcu et al. 2016: 7).

A carbon tax is a tax on fossil fuels related to their carbon content, designed to encourage energy users to substitute fossil fuels with low-carbon energy sources such as renewables and cleaner fossil fuels. Since it would raise overall energy prices, it should also result in reduced energy usage. This could happen through reducing consumption (e.g., heating one's house less) but it could also incentivise improvements in energy efficiency, in the residential sector (such as through retrofitting insulation in housing) or in industry (through innovation in product design to low-carbon alternatives). It is estimated that carbon taxes of some form are now levied in 40 countries and 16 other states or provinces around the world (Carl and Fedor 2016: 51). These include the Nordic countries (Norway, Denmark, Sweden, Finland and Iceland) which were the first to introduce them in the early 1990s, Switzerland, the UK, Ireland, France and the Netherlands in Europe, as well as Costa Rica, India, Japan and Mexico. An Australian tax, introduced in 2012 was abolished in 2014 following a change of government. Chile has announced a modest tax, the first in South America, to be introduced in 2018. In 2008, the Canadian province of British Columbia instituted North America's first comprehensive and substantial carbon tax (see Box 8.1).

Comparing emissions trading with carbon taxes, Carl and Fedor find that, using 2013 figures, carbon taxes collected a total of $21.7bn whereas emissions trading (or cap-and-trade schemes as they are also called) generated only $6.57bn. However, while 70% of the latter revenue was dedicated to 'green spending' (energy efficiency measures or renewable energies), 72% of the revenue from carbon taxes went either to general state funds or was used to fund tax breaks (Carl and Fedor 2016: 60). Carbon taxes appeared to fall from favour around the mid to late 2000s as attention focused on the greater political feasibility of cap-and-trade systems. However, the volatility of cap-and-trade saw a return to carbon taxes in the late 2000s, more for their stability and ability to raise revenue for hard-pressed public revenues (such as in recession-hit Ireland and Iceland) than for their ability to mitigate GHG emissions. While new schemes are being regularly established, total revenues raised remain relatively modest (10s of billions of dollars annually). 'The variation across a multitude of young systems, in both approach and outcome, suggests a lack of strong public or political norms to guide policymakers and constituents through the lawmaking process', they add (ibid.).

However, the literature also emphasises that carbon pricing alone is insufficient. Sandén and Azar (2005) argue that carbon cap-and-trade is important for diffusion—'picking technologies from the shelf'—but insufficient for innovation—'replenishing the shelf'. Foxon (2003) emphasises the interaction of environmental and knowledge market failures, arguing that this creates 'systemic' obstacles that require government action beyond simply fixing the two market failures (of climate damages and technology spillovers) independently. There is therefore general consensus in the literature that, while emission reduction (including pricing) mechanisms are a necessary component for delivering such innovation, they are not sufficient: efficient innovation requires even more government action (Barker et al. 2007: 661–662).

Box 8.1: British Columbia: Achieving Emissions Goals
British Columbia's stand-alone carbon tax, introduced in 2008, covered about three-quarters of all emissions sources in the province at a levy rate higher than most ETS systems. Viewed by Murray and Rivers as a 'grand policy experiment', the tax was reviewed by the provincial government in 2013 and judged a success not requiring any major changes (Murray and Rivers 2015: 675). Introduced at C$10 a ton, the price rose by C$5 a ton until it reached C$30 in 2012 where it remains. This is the highest in North America, twice as high as in Alberta.

In their assessment, Murray and Rivers find that the tax has resulted in reductions in GHG emissions of between 5% and 15%, with 'no statistically significant effect at all on net growth'. Fears that the tax would fall disproportionately on low-income households led to a proportion of its revenues being dedicated to credits and tax cuts for these households which seem to have mitigated any regressive impacts though this may have worsened as the tax rate increased. Furthermore, public support for the tax increased over time, though not in all sectors of the population.

However, Murray and Rivers warn of emissions leakage in that 'at least some of the reductions in emissions observed in British Columbia are likely to be associated with increases in emissions elsewhere' (ibid.: 682), something yet to be quantified. This is in line with the view that a domestic carbon tax 'could encourage production of carbon-intensive goods to shift to low-carbon tax jurisdictions' (Marron and Toder 2014: 564).

Implementing Climate Capitalism

Mechanisms such as cap-and-trade or carbon taxes are but the most visible manifestations of a wider approach to addressing the climate crisis within the framework of capitalism. Newell and Paterson trace the growing awareness of climate change among financial actors since the early 1990s. Not surprisingly, large insurance companies were among the first to view climate as a risk to their viability. While withdrawing cover in areas considered vulnerable to climate change was one response, another was the development of new financial instruments such as 'weather derivatives' or 'catastrophe bonds', illustrating the power of finance to turn threats into opportunities to make a profit. As shareholders woke up to the new realities posed by the climate, pressure mounted on companies to disclose their GHG emissions which began to be factored into the investment decisions, since such emissions could hit their profits as governmental regulations to limit emissions became more commonplace. A leading company in financial services, J.P. Morgan began to assess the financial risks of emissions in loan evaluations. As the Kyoto Protocol came into force and emissions trading systems were set up, so too were a series of differentiated carbon finance products rolled out, such as derivative markets, information-diffusing mechanisms and even a credit-rating instrument, 'IDEAcarbon' (Newell and Paterson 2010: 60–77). This shift in seeing climate as an opportunity rather than as a threat has helped to mobilise a big increase in investments in renewable energy that went from $62 billion in 2004 to $329 billion in 2015 (Bloomberg 2016). Therefore, Sapinski notes that many authors see the energy and financial sectors as having played 'a crucial role in the emergence and functioning of climate capitalism' (Sapinski 2016: 102).

In 2014, the Global Commission on the Economy and Climate, which includes former presidents (Felipe Calderón of Mexico and Ricardo Lagos of Chile), former prime ministers (Helen Clark of New Zealand, Jens Stoltenberg of Norway and Luísa Diogo of Mozambique), Angel Gurría, secretary general of the Organisation for Economic Co-operation and Development (OECD), Nicholas Stern who wrote the Stern Report on the economic costs of climate change for the UK government, and a number of high-level bankers, issued its first report entitled *Better Growth, Better Climate: The New Climate Economy Report* (GCEC 2014). It argues that the next 15 years offer a critical opportunity to undertake 'a deep structural transformation' of the global economy, 'to build lasting economic growth, at the same time as reducing the immense risks of climate change. ... Future economic growth does not have to copy the high-carbon, unevenly distributed model of the past' (GCEC 2014: 8). Instead 'there is now huge potential to invest in greater efficiency, structural transformation and technological change in three key systems of the economy', cities, land use and energy (ibid.). Two reports followed: the first in July 2015 was *Seizing the Global Opportunity: Partnerships for Better Growth and a Better Climate* (GCEC 2015) focusing on how international and multistakeholder co-operation could catalyse sustainable economic growth while also reducing emissions; the second in October 2016, *The Sustainable Infrastructure Imperative: Financing for Better Growth and Development* (GCEC 2016), identifies the barriers to financing sustainable infrastructure and proposes measures to overcome them. The project is now launching research on specific countries and sectors.

This series of reports is a major attempt to galvanise global action to shift decisively to a low-carbon society while at the same time reigniting global growth and meeting the Sustainable Development Goals (SDGs). It is nothing if not ambitious: 'The next 2–3 years will be crucial in bringing about a fundamental change of direction', it states in the 2016 report (GCEC 2016: 2). At the heart of how this can be done is investment in 'climate-smart, resilient infrastructure': 'Ensuring infrastructure is built to deliver sustainability is the only way to meet the global goals ... and to guarantee long-term, inclusive and resilient growth'. This includes what the report calls 'traditional infrastructure' such as energy, public transport, buildings, water supply and sanitation, but also 'natural infrastructure' including forest landscapes, wetlands and watershed protection (ibid.: 4). It repeats the need for investment of around US$90 trillion in global infrastructure over the next 15 years, an increase from the estimated $3–4 trillion annually at the moment to about $6 trillion a year, two-thirds of it to be spent in the global South. This will require a combination of public and private finance, as well as a catalytic role for multilateral regional and bilateral development finance institutions as well as national development banks. Among its first priorities is tackling 'fundamental price distortions'—including phasing out public subsidies for fossil fuels which amounted to around $550 billion globally in 2014, which exceeds public subsidy for renewable energy a number of times over. It says that 'strong, effective and rising carbon prices' are also a 'necessary condition for inclusive and low-carbon growth' (GCEC 2016: 6). Furthermore pricing needs to include 'the social costs of externalities, for example the costs of air pollution from fossil fuel use as well as of congestion from urban vehicle use' (ibid.: 7). The financial system needs to be transformed to reduce the cost of capital, enable catalytic finance from development finance institutions and to accelerate the greening of the financial system. Noting that the green bond market reached $42 billion in 2015, it recommends mobilising green finance 'to prioritise and value more sustainable long-term investments over a narrow focus on short-term gains' (ibid.: 11). Finally, it urges increased investment in clean technology research, and development and action to reduce the costs of more sustainable technologies. The Global Commission says that 'this is the only sustainable, long-term growth path on offer, bringing with it a means to increase living standards, promote inclusion and reduce poverty' (ibid.: 12).

While the New Climate Economy reports embody the essential features of climate capitalism as outlined by Newell and Paterson, they combine these market-based mechanisms with a positive role for the public sector in driving change. To this extent, they constitute a proposal to develop climate capitalism in a more Keynesian direction, mildly echoing proposals for a Green New Deal that emerged as the global recession hit in 2008 (see Box 8.2). However, the New Climate Economy reports owe more to the concept of the 'green economy', which initially emerged as 'green growth' pioneered by South Korea (see Chap. 7), and was developed in three major reports, by the UNEP (2011a), the OECD (2011) and the World Bank (2012). Central to the concept is attaching a more appropriate monetary value to 'natural capital'[1] through carbon taxes and trading permits as well as the removal of fossil-fuel subsidies, but it also emphasises the need for better information, awareness raising and the enforcement of tougher standards to reduce behavioural resistance to the changes needed.[2] Through substantial investment in R&D, it also holds out the hope of developing new technologies that help decarbonise the economy. In assessing the potential of the concept of the green economy, Turok and Borel-Saladin (2013) emphasise its 'positive vision of the future' that can help to inspire change rather than resistance in citizens and decision-makers. 'The emphasis is on pursuing the combined benefits of interactions between the economy and the environment, rather than accepting trade-offs and compromises', they write (Turok and Borel-Saladin 2013: 289). But they state that there are 'different versions of the green economy, each implying different levels of intervention and different outcomes', ranging from 'minor incremental reforms to major restructuring and transformation of the system'. None of the three major reports referred to here address this question explicitly, they add (ibid.: 291). Instead, they go on to identify the core logic of the concept:

> There is a technocratic slant to these reports which verges on assuming that if natural resources are priced correctly, the economy will green itself. There should be operating-cost savings from some green technologies and more efficient systems of production and distribution, but these do not mean that the green economy will emerge automatically. In the face of considerable, inertia, vested interests and investments already made, it is likely that co-ordinated political action will be required to achieve the systemic changes envisaged. (ibid.: 292)

One of the areas in which strong political action will be required is in the need for 'a substantial transfer of resources to developing nations', without which most will struggle to be able to invest in the transition to a green economy (ibid.: 292). Strong political action will also be required to ensure that the transition happens in ways that reduce poverty and ensure equity. Since it is based on continuing economic growth, a third challenge for climate capitalism relates to the urgent need to decouple such growth from GHG emissions. Yet, as Jänicke points out, 'so far there has been only a relative decoupling of economic growth and resource consumption in some advanced countries and the overall relief has been neutralised through rebound effects' since technology efficiency and saving stimulated increased consumption (Jänicke 2012: 19). These then form three major challenges to climate capitalism: the challenge of finance, the challenge of governance and the challenge of decoupling. Each will be examined before considering Turok and Borel-Saladin's fundamental questions: 'Will the scale of change from "business as usual" be sufficient to prevent excessive global warming and other environmental catastrophes, bearing in mind continuing population growth and pressures to increase consumption? Can a new sustainable development path be engineered by manipulating resource prices and stimulating new technologies? Or does the underlying market-based, short-term, growth-oriented paradigm of the global economy need to be replaced?' (Turok and Borel-Saladin 2013: 291).

Box 8.2: Greening Roosevelt's New Deal

Drawing inspiration from Franklin D. Roosevelt's response to the Great Crash of 1929, a group of progressive economists linked to the New Economics Foundation in Britain launched a proposal in July 2008 for a Green New Deal as the global recession hit (Green New Deal Group 2008). Identifying 'the scale of the menace posed to the natural world, the global economy and all our livelihoods by a triple crunch' (ibid.: 6), the document proposes 'the transformation of national economies and the global economy' (ibid.: 23) to address the financial crunch, the climate crunch and the energy crunch.

It would entail re-regulating finance and taxation coupled with state intervention to mobilise higher public and private expenditure, fostering employment and demand through investment in infrastructure and 'targeting environmental projects that will dramatically cut fossil fuel use and hence help to tackle climate change and peak oil' (ibid.: 23). To achieve this, it promotes '"joined-up thinking" about the four systems that dominate our world: the market, the state, civil society and the ecosystem' (ibid.: 6).

The idea of a Green New Deal was taken up by the UN Environment Programme (UNEP) in its Global Green New Deal document proposing investment to create jobs in 'green' industries (UNEP 2009), by the European Green Party and by the Green Party candidate in the US presidential elections of 2012 and 2016, Jill Stein. However, it was soon overtaken by the emergence of a more substantive set of proposals from a number of international organisations, centred on the concept of the 'green economy'.

Challenges I: Finance

The New Climate Economy project depends crucially on mobilising finance of around $90 trillion over the next 15 years and the Commission says in its third report that 'this is an especially opportune moment' to make such choices due to low interest rates and rapid technological change (GCEC 2016: 4). Yet, far more pertinent to the possibility of realising this scale of investment is the nature and capacity of today's financial system which is overlooked by the New Climate Economy reports. Often called 'financialisation', this refers to a series of reforms to the global financial system beginning with the announcement by President Nixon in 1971 that the US was no longer linking the value of the dollar to gold which had underpinned global financial stability for four decades. Various restrictions, most particularly the separation of investment from commercial banking that had been introduced in the Glass-Steagall Act in 1933, were progressively relaxed in the 1980s and 1990s. These included the 'Big Bang' deregulation of the British financial sector in 1986 and the final repeal of the Glass-Steagall Act in the US in 1999 resulting in a huge growth in the financial system but also a new volatility. As Dietz and O'Neill point out, following deregulation in Britain, the money supply began to grow much faster than GDP for the first time. As they explain: 'In recent years, the money supply has become almost completely detached from the real economy, as new financial instruments have allowed banks to pump more and more money into the economy. The disconnect has caused much of the economic and financial instability in the world today' (Dietz and O'Neill 2013: 102). Volumes of trading in financial markets reached absurd levels, writes Kay, and 'increased rather than diversified the risks to which the global economy is exposed' (Kay 2016: 298). The collapse of many banks in 2008 exposed the volatility of the system but as yet has failed to motivate its fundamental reform.

A number of features make this liberalised financial system poorly equipped to supply the type and levels of finance required by the New Climate Economy project. Firstly, as Kay puts it: 'Financialisation has forced all businesses to take a more short-term view' (Kay 2016: 255). Yet, it is precisely the opposite that is required to green the economy, namely 'higher up-front financing, with the savings and other benefits accruing later' (GCEC 2016: 4). Newell writes: 'Private financial investment is clearly driven by the prospect of rapid returns from attractive accumulation sites in the global economy, not by where finance is most needed for sustainable development' (Newell 2012: 115). Flows of private unregulated capital, which dwarf the amounts of public finance available, tend rather 'to undo, bypass and overwhelm positive and incremental gains achieved through use of public monies' (ibid.: 131). Furthermore, the multilateral development finance institutions, such as the World Bank, which are seen by the New Climate Economy project as playing a 'catalytic role' in shaping and directing action to build the sustainable infrastructure of the twenty-first century, have come in for a lot of criticism since lending is often based on the commercial interests of Northern firms or on political expediency (Newell 2012: 131). Another feature of today's financial system is its huge levels of debt (Box 8.3). Newell argues that a different model of environmental governance is required for private financial actors, embracing both regulation and accountability, if they are to provide the long-term investments in infrastructure for sustainable development required for a successful climate capitalism. Yet, as he writes, 'the record of states, alongside multilateral development actors such as the World Bank … does not offer solid grounds for believing that the governance of finance for sustainable development will be strengthened any time soon' (Newell 2012: 143). The sorts of reforms needed would require a fundamental overhaul of the whole global financial system; despite facing collapse in 2008, no substantial reform took place then (Kay 2016: 300).

Box 8.3: IMF Warns of Debt Timebomb

Debt as a proportion of GDP in the global economy has never been higher, warned the IMF in late 2016. In its first comprehensive examination of the issue, it estimated global debt at $152 trillion or about 225% of global output. As Elliott reported: 'The IMF says there is a debt timebomb out there. And the ticking is getting louder' (Elliott 2016).

What distinguishes today's situation from other periods of high debt, such as after the Second World War, is that since 2009 both growth and inflation have been low by historic standards. Thus the conditions don't exist for the global economy to grow out of the problem as happened in the post-War period. Instead, the fear is that the global economy could be in the sort of debt-deflation spiral that happened in the 1930s.

So, within today's global financial system, how much funding is actually flowing to deal with climate change? The most authoritative evidence comes from the UNFCCC's biannual assessment of global climate finance flows. The 2016 report identified climate-related 'total global climate finance flows'[3] of USD$880 billion in 2013 and $930 billion in 2014 (UNFCCC 2016: 56). The largest component of this finance in all countries (developed and developing) is private investment in renewable energy, which amounted to $265 billion in 2014 and domestic public finance expenditures of some $192 billion. However, the UNFCCC adds: 'This is a significant amount, but it is relatively small in the context of wider trends in global investment. For example, while investment in clean energy is rising, volumes of finance for high carbon energy in all countries remain considerably higher. Infrastructure and assets are at risk from the impacts of climate change, with serious potential consequences for the global economy' (UNFCCC 2016: 9). The IEA has estimated that just implementing the energy actions in the INDCs will cost $16.5 trillion from 2015 to 2030, or more than $1.1 trillion per year, while every year $1.6 trillion is invested in fossil fuel energy (UNFCCC 2016: 92).[4] While the costs of renewable energy are coming down, and the flows of climate finance are indeed increasing, the climate finance estimates do not tally with needs. They are currently insufficient to meet the cost of energy actions envisaged in INDCs,[5] which themselves are insufficient to meet even the 2°C target. They are also insufficient to meet the additional financing needs required for adaptation to the climate change that cannot be avoided. Standing in contrast to the still higher annual investments in fossil fuels, it is clear that while progress has been made, that fossil-fuelled 'brown growth' is still *du jour*, and there is major inertia in aligning financial patterns to the scientifically and politically agreed targets for limiting climate change.

Challenges II: Governance

Despite its focus on how to bring climate to the heart of decision-making, the New Climate Economy approach and the concept of the green economy which it mirrors are essentially very technocratic, hoping for a massive shift in investment and for major technological deployment. While containing many useful ideas on resource efficiency, infrastructure investment and innovations in technologies, business models and social practices, it is silent on the type of state or the nature of political engagement that might deliver these necessary transformations. There are, of course, many mentions of what governments (including municipal governments) should do but there is no recognition that the radical transformations needed also require the transformation of our states and our governance processes if they are to be achieved. By contrast, the German Advisory Council on Global Change (WBGU) in its report on a social contract for sustainability (WBGU 2011) recognises in its first priority action the need for a proactive and democratic state, as well as the importance of policy-making processes in some of its other action points. What distinguishes the WBGU report is the extensive attention it devotes to addressing the challenges to governance that the transformation of cities, land use and energy requires.

The report recognises that 'any illusion that a regime change on a global scale can be purely technocratically organised and steered top-down must be avoided' (WBGU 2011: 175). Instead, it focuses on 'the opportunities and limits of governance', identifying the barriers that impede the transformation process, and examining approaches to solving the problems encountered. Thus it examines what it calls 'transformation governance', including the altered role of the state at global, national and local levels requiring 'new normative foundations, improved instruments and unusual approaches on all levels of governance' (ibid.: 176). Indeed, it is recognised that the strengths of modern statehood such as 'gaining time through compromise, integration of well-organised interests into the political decision-making process (neo-corporatism) and a well-meaning balancing of these on the part of a moderating state', have now become obstacles to the transformations needed. Instead, to overcome what it calls 'short-term orientation and the politics of delay' (ibid.: 189) a 'new statehood' is advocated. This is described as follows:

> In political terms, the major transformation process challenges are the acceleration of politico-administrative procedures and processes, improvements with regard to the implementation of a long-term orientation in policy-making, the resolute overcoming of path dependencies, the empowerment and involvement of the civil society, and a historically unparalleled expansion of international cooperation. (ibid.: 203)

Central to this transformation of governance is that states return to playing a more active role in regulating the economy and society. This, then, echoes the 'climate Keynesianism' outlined by Newell and Paterson

(Newell and Paterson 2010: 172–178), combining stronger state regulation of markets with action for wider social transformation both towards decarbonisation and towards greater distributional equality. But the WBGU also recognises that 'top-down government planning is an illusion': 'The state itself does not know the best options, but is tasked with activating both corporate and civil society, and politico-administrative system potentials, whilst also refraining from restricting itself to the purely moderating and remedial role that is typical for pluralistic negotiation democracies' (ibid.: 203). So, more state interventionism needs to be balanced by more citizen involvement in the form of a new social contract.

The new 'proactive and enabling state', needed for the transition to a low-carbon society, needs to balance two principles that have for long been seen as opposed to one another. These are, on the one hand, the empowerment of the state to actively determine priorities (a task which, over recent decades, has been handed over to private market actors) and, on the other, 'providing citizens with more extensive opportunities to have a voice, to get more involved in decision-making processes, and to take on a more active role in politics' (ibid.: 209). Yet, in the context of moving to forms of development that respect the limits of global ecosystems, the actions that the state is required to take to ensure that this happens must be consistent with the interests of citizens for a sustainable future. A major challenge therefore is the aligning of these objectives of the state and of citizens so that both are seen to work collaboratively. As the title of one section in the WBGU report puts it succinctly: 'Transformation impediments and barriers: It's politics, stupid!' (ibid.: 188).

Though the WBGU does not use the term, their approach focuses on the need for a new political economy model, whereby the state plays a much more transformative role in the economy and society but is itself held in check by much firmer and more deliberative mechanisms of citizen engagement. This would constitute a very different form of climate capitalism to today's mainstream form, and indeed, could lead beyond capitalism into an ecosocialism (see Chap. 9).

Challenges III: Decoupling

Concluding his study of the corporate-funded climate and environmental policy groups (CEPGs) that promote climate capitalism, Sapinski sums up their objective as being 'to establish new bases for accumulation within a broadly neoliberal order' through diverting 'financial flows from the oil and coal sectors and GHG emitting electricity production, and to redirect them towards supporting the ecological modernisation of capitalist production processes'. This 'new regime of accumulation' would partially internalise certain environmental externalities and 'decouple economic growth from the growth of GHG emissions' (Sapinski 2016: 104–105). The real test of today's dominant form of climate capitalism, then, is the extent to which it can decouple growth from emissions. As Gupta writes: 'Without decoupling, continuing and increasing economic growth in developed and developing countries would come with ever-increasing environmental pressures, unavoidably destroying the carrying capacity of ecosystems with corresponding detrimental effects on the environment and societies' (Gupta 2015: 510).

In his discussion of what he labels 'the myth of decoupling' (Jackson 2009: 68), Jackson makes the vital distinction between relative and absolute decoupling. The former refers to a situation in which emissions can decline per unit of GDP but if GDP continues to grow, so do emissions in an absolute sense. Since we need to reduce emissions by up to 85% to 90% by 2050 to keep warming within 2°C, an absolute decoupling of growth from emissions is required. Examining evidence up to the mid-2000s, Jackson found that global energy intensity, namely the amount of primary energy needed to produce each unit of economic output, has fallen by about one-third since the early 1970s. But these gains are most evident in the advanced economies; outside these, energy intensity has been increasing. Overall, global carbon intensity declined by almost a quarter between 1980 and 2006 but the declining trend faltered towards the end of that period. Evidence of absolute decoupling is harder to find, writes Jackson, as carbon dioxide emissions were almost 40% higher in the mid-2000s than in 1990 (ibid.: 67–86). Even the relative stabilisation seen in some developed countries disappears since these data fail to take into account the emissions embedded in imported goods; once these are included the emissions of these countries are seen to have risen rather than fallen. Indeed, studying the emissions of the three largest exporting countries in the world, the US, Germany and China, Gurtu et al. (2016) found that, once outsourced emissions are included in national figures, 2010 emissions in the US and Germany jumped by 18% and 20% respectively, but emissions in China dropped by 6%.

and the use of markets are proving successful in reducing emissions is one thing, he writes, but it is quite another to claim that 'capital's growth alignments are met through low-carbon forms of development' since 'the still small percentage of economic flows currently constituted by clean energy ... continue to be overwhelmed by fossil fuel use' (Newell 2012: 151). Beyond the technical possibilities that tend to dominate the discourse on climate change, Newell highlights the realities of today's capitalist order. Any conclusions about how likely is climate capitalism to get us to a post-carbon society, therefore, need to deal with these realities.

While a climate capitalism based on the characteristics of neoliberalism is the dominant model today, other forms also exist. The statist capitalism of China is one example while the more social democratic models of Costa Rica and Uruguay have also been highlighted. As Newell and Paterson (2010) recognise, there are different versions of climate capitalism. While some seem more successful than others in reducing emissions and planning for adaption to climate change, they have characteristics in common that mark out the limits of transformation possible within the confines of a model dominated by the need for capital accumulation. The principal conclusion that can be drawn from the analysis of climate capitalism in this chapter is that it faces an overwhelming challenge in addressing the need for substantial reductions in the consumption of energy and resources (see Box 8.5). Is this going to be possible in a system driven by the needs of capital accumulation?

There are two sides to the question. The most obvious one is the need to reduce consumption but this needs to be done equitably, so that the benefits and costs are more equally shared (see Box 9.3). The field of 'sustainable consumption and production' (SCP) has, since the early 1990s, studied the requirements of altering consumption patterns in the light of climate change. In a survey of the field, Pogutz and Micale conclude that, even though SCP has for decades been considered a critical issue for policy makers and civil society, changes achieved are not enough 'to effectively address destructive consumption trends, fostered by materialism and by the access of an increasing number of developing countries to global markets and to the material-intensive attitudes inspired by our model of economic development'. The demanding challenge now, they write, is for business and governments 'to cooperate with developing countries in constructing new economies vastly different from our societal model' (Pogutz and Micale 2011: 47). In a similar vein, focusing on the measures needed to make moves towards lower consumption patterns more equitable, Gough outlines some of the measures necessary:

> Together, the now very unequal distribution of personal consumption would need to be addressed, via socialized consumption, taxation, public transfers, and 'pre-distributive' measures such as minimum wages, maximum rewards, and trade-union rights. These are radical shifts that would challenge dominant interests and ideas, for example, 'consumer sovereignty' and unquestioned economic growth. (Gough 2016: 42)

The need for changes far more radical than anything achieved up to now brings another dimension of capitalism into focus, usually neglected in analyses of climate capitalism but referred to by the WBGU report in its emphasis on the importance of politics. This, as Frase puts it, is 'capitalism as a system of class power, with a ruling elite that will try to preserve itself into any possible future'. Consistent with the critique already developed in Chaps. 1 and 2 of this book, he writes that the role technology plays needs to be considered in this context: 'Technological developments give a context for social transformations, but they never determine them directly; change is always mediated by the power struggles between organized masses of people' (Frase 2016: 30). Central to Frase's analysis is that technology has the potential to get us to a post-carbon society but only in a radically changed society in which power is fundamentally re-distributed through social struggle.

It is for this reason that Naomi Klein sees a basic conflict between capitalism and the climate. What she calls 'the fetish of centrism', the gradual, incremental options through which climate capitalism addresses the challenges, is no longer adequate. The demands of profit-making trump the climate imperative, she concludes, realising that 'the oligarchs who were minted by the era of deregulation and mass privatization are not, in fact, going to use their vast wealth to save the world on our behalf' (Klein 2014: 255). She adds that, 'underneath all of this is a real truth we have been avoiding: climate change isn't an "issue" to add to the list of things to worry about, next to health care and taxes. It is a civilizational wake-up call. A powerful message—spoken in the language of fires, floods, droughts and extinctions—telling us that we need an entirely new economic model and a new way of sharing this planet' (Klein 2014: 22, 25). Lorek and Spangenberg suggest one link to where this new model may be in gestation when they identify the alternative required as being 'strong sustainable consumption' focusing not primarily on technology 'but on affluence, the level and patterns of resource consumption or the physical size of the economy, thus providing a link to the ongoing de-growth discourse' (Lorek and Spangenberg 2013: 35). This is examined in Chap. 9.

Box 8.5: Bridging the Gap: Gambling or Degrowth?
Recognising that a gap exists between what is possible using the most ambitious renewable energy, energy efficiency and low-carbon land-use measures, and the scale of emissions reductions required, Wiseman and Alexander (2017) identify the three options available to bridge the gap:

1. Gambling on mitigation technologies which in most cases are far from proven;
2. Gambling that adaptation to warming of 4°C or more is possible;
3. Reducing energy and resource consumption by the better-off sectors of humanity.

Only the last of these can address other global ecological challenges such as ocean acidification, and the collapse of biodiversity, they add. They conclude that it is necessary to include 'planned and equitable reductions in the consumption of energy and resources in debates about the suite of actions required to meet global climate change and planetary boundary challenges' (Wiseman and Alexander 2017: 101).

A final point to be made about climate capitalism relates to the future of capitalism itself. Five distinguished scholars of international systems came together to write a book on capitalism's future because, despite their many differences, they all agree that 'something big looms on the horizon: a structural crisis much bigger than the recent Great Recession, which might in retrospect seem only a prologue to a period of deeper troubles and transformations' (Wallerstein et al. 2013: 1–2). The question which is the title of the book 'Does Capitalism Have a Future?' emerges not primarily because of climate change but because of other deep crises of the system. German sociologist Wolfgang Streeck goes further in a subsequent book, asking 'How Will Capitalism End?'. He identifies five 'systemic disorders' of contemporary capitalism: 'stagnation, oligarchic redistribution, the plundering of the public domain, corruption and global anarchy' that pose a question mark over the future of the system itself (Streeck 2016: 28). Paul Mason, former economics editor of Channel 4 News sees the rise of information technology as disrupting basic institutions of capitalism such as prices, ownership and wages. With shareable information goods the basic law of capitalist economics is turned on its head: instead of scarcity we have abundance so that supply and demand become irrelevant. This undermines the normal operation of the price mechanism and 'has revolutionary implications for everything' (Mason 2015: 120): 'The technologies we've created are not compatible with capitalism—not in its present form and maybe not in any form. Once capitalism can no longer adapt to technological change, postcapitalism becomes necessary' (ibid.: xiii). Whether the focus is on crisis or potential, it is clear that for some major analysts the future of capitalism itself is at stake.

On this reading then, the requirements of transitioning to a post-carbon society by 2050 and the requirements of information technology point to the inability of capitalism to offer adequate pathways to a better future. A major conclusion of this chapter must be that while a post-carbon future is, in Mason's terms, technologically feasible and economically rational, 'what stands in the way is the market', namely the institutionalised vested interests of a capitalist elite. 'The attempt to create a non-market economy and a low-carbon system are clearly interdependent', he writes, the only question is how it is going to happen (251–252). The conclusions we can draw from this chapter is that it will require a new development model in which the state has far greater power to direct market forces, and civil society has a much more active role to play to ensure socially just and sustainable outcomes. Chapter 9 examines such an alternative model.

Notes

1. 'Natural Capital' was defined by the World Forum on natural capital in 2015 'as the world's stocks of natural assets which include geology, soil, air, water and all living things. It is from this Natural Capital that humans derive a wide range of services, often called "ecosystem services", which make human life possible.' Economic definitions of 'capital' can be problematic where they give the impression that different forms of capital are fully substitutable. This is plainly not possible for many forms of natural capital on which we rely for our existence.

2. Individual behaviour change through provision of information and awareness raising is a common prescription from the rational actor model of neoclassical economics, but it is one that is encountering significant real world problems in delivering change, see Fleurbaey et al. (2014).
3. Where total global climate finance flows include the following categories: international public climate finance, private and public investment in renewables, private investment in energy efficiency, private investment in sustainable transport, climate-related land use, and adaptation and domestic climate-related public investment (UNFCCC 2016: 53).
4. The New Climate Economy report has estimated seperately that global infrastructure requirements in the case of a high-carbon economy (across transport, energy, water systems and cities), require investment of around $90 trillion over the next 15 years, or an average of $6 trillion per year (GCEC 2015).
5. The INDCs are the Intended Nationally Determined Contributions which all countries submit to the UNFCCC as their policies towards low-carbon transition. The IEA estimates of the 'energy actions' are the aggregated sum of all policy actions from all countries to reduce emissions from energy as detailed in INDCs submitted to the UNFCCC.
6. Perhaps the most widely known paradox in environmental economics, the Jevons paradox refers to a situation when, instead of technological progress leading to a decline in consumption, the opposite occurs. The English economist William Stanley Jevons in 1865 observed the paradox when he found that increased cost-saving efficiencies in the production of coal, resulted in its greater use in a wide range of industries. The concept has since been extended to the use of any resource, including fossil fuels. While widely referred to, doubts have also been raised about how strong it is (see Gillingham et al. 2013).

References

Barker, T., I. Bashmakov, A. Alharthi, M. Amann, L. Cifuentes, J. Drexhage, M. Duan, et al. 2007. Mitigation from a Cross-Sectoral Perspective. In *Climate Change 2007: Mitigation. Contribution of Working Group III to the Fourth Assessment Report of the Intergovernmental Panel on Climate Change*, ed. B. Metz, O.R. Davidson, P.R. Bosch, R. Dave, and L.A. Meyer, 619–690. Cambridge and New York: Cambridge University Press.

Bloomberg. 2016. Clean Energy Defies Fossil Fuel Price Crash to Attract Record $329bn Global Investment in 2015. Accessed 2 December 2016. data.bloomberglp.com

Carl, Jeremy, and David Fedor. 2016. Tracking Global Carbon Revenues: A Survey of Carbon Taxes Versus Cap-and-trade in the Real World. *Energy Policy* 96: 50–77.

Csereklyei, Zsuzsanna, and David I. Stern. 2015. Global Energy Use: Decoupling or Convergence? *Energy Economics* 51: 633–641.

Dietz, Rob, and Dan O'Neill. 2013. *Enough is Enough: Building a Sustainable Economy in a World of Finite Resources*. London: Routledge.

EC. 2016. Emissions Trading: Slight Decrease in Emissions in 2015. Accessed 22 November 2016. https://ec.europa.eu/clima/news/articles/news_2016052001_en.htm

Elliott, Larry. 2016. Governments Must Heed IMF Warning of $152tr Global Debt Timebomb. *The Guardian*, 5 October 2016.

Fleurbaey, M., S. Kartha, S. Bolwig, Y.L. Chee, Y. Chen, E. Corbera, F. Lecocq, et al. 2014. Sustainable Development and Equity. In *Climate Change 2014: Mitigation of Climate Change. Contribution of Working Group III to the Fifth Assessment Report of the Intergovernmental Panel on Climate Change*, ed. O. Edenhofer, R. Pichs-Madruga, Y. Sokona, E. Farahani, S. Kadner, K. Seyboth, A. Adler, I. Baum, S. Brunner, P. Eickemeier, B. Kriemann, J. Savolainen, S. Schlömer, C. von Stechow, T. Zwickel, and J.C. Minx. Cambridge: Cambridge University Press.

Foxon, T.J. 2003. *Inducing Innovation for a Low-Carbon Future: Drivers, Barriers and Policies*. London: The Carbon Trust.

Frase, Peter. 2016. *Four Futures: Life after Capitalism*. London: Verso.

Gillingham, K., M. Kotchen, D. Rapson, and G. Wagner. 2013. The Rebound Effect is Over-played. *Nature* 493: 475–476.

Global Commission on the Economy and Climate. 2014. *Better Growth, Better Climate: The New Climate Economy Report*. Washington, DC: New Climate Economy.

———. 2015. *Seizing the Global Opportunity: Partnerships for Better Growth and a Better Climate*. Washington, DC: New Climate Economy.

———. 2016. *The Sustainable Infrastructure Imperative: Financing for Better Growth and Development*. Washington, DC: New Climate Economy.

Gough, Ian. 2016. Welfare States and Environmental States: A Comparative Analysis. *Environmental Politics* 25 (1): 24–47.

Green New Deal Group. 2008. *A Green New Deal*. London: NEF.

Gupta, Shilpi. 2015. Decoupling: A Step Toward Sustainable Development with Reference to OECD Countries. *International Journal of Sustainable Development & World Ecology* 22 (6): 510–519.

Gurtu, Amulya, Cory Searcy, and Mohamad Y. Jaber. 2016. A Framework for Reducing Global Manufacturing Emissions. *Journal of Environment & Development* 25 (2): 159–190.

Jackson, Tim. 2009. *Prosperity without Growth: Economics for a Finite Planet*. London: Earthscan.

Jänicke, Martin. 2012. "Green Growth": From a Growing Eco-Industry to Economic Sustainability. *Energy Policy* 48: 13–21.

Kay, John. 2016. *Other People's Money: Masters of the Universe or Servants of the People?* London: Profile Books.
Klein, Naomi. 2014. *This Changes Everything: Capitalism vs. the Climate*. London: Penguin.
Laing, Tim, and Michael Mehling. 2013. International Experience with Emissions Trading. *Climate Strategies*. www.climatestrategies.org
Lorek, Sylvia, and Joachim H. Spangenberg. 2013. Sustainable Consumption within a Sustainable Economy—Beyond Green Growth and Green Economies. *Journal of Cleaner Production* 63: 33–44.
Marcu, Andrei, Milan Elkerbout, and Wijnand Stoefs. 2016. 2016 State of the EU ETS Report. *Carbon Market Forum*. Brussels: Centre for European Policy Studies.
Marron, Donald B., and Eric J. Toder. 2014. Tax Policy Issues in Designing a Carbon Tax. *American Economic Review: Papers & Proceedings 2014* 104 (5): 563–568.
Mason, Paul. 2015. *Postcapitalism: A Guide to our Future*. London: Penguin.
Meadows, Donella H., Dennis L. Meadows, Jorgen Randers, and William W. Behrens III. 1972. *The Limits to Growth*. New York: Universe Books.
Meadows, Donella, Jorgen Randers, and Dennis Meadows. 1992. *Beyond the Limits*. Post Mills, VT: Chelsea Green Publishing Company.
———. 2004. *Limits to Growth: The 30-Year Update*. White River Junction, VT: Chelsea Green Publishing Company.
Murray, Brian, and Nicholas Rivers. 2015. British Columbia's Revenue-Neutral Carbon Tax: A Review of the Latest "Grand Experiment" in Environmental Policy. *Energy Policy* 86: 674–683.
Newell, Peter. 2012. *Globalization and the Environment: Capitalism, Ecology and Power*. Cambridge: Polity.
Newell, Peter, and Matthew Paterson. 2010. *Climate Capitalism: Global Warming and the Transformation of the Global Economy*. Cambridge: Cambridge University Press.
OECD. 2011. *Towards Green Growth*. Paris: OECD.
Pogutz, Stefano, and Valerio Micale. 2011. Sustainable Consumption and Production: An Effort to Reconcile the Determinants of Environmental Impact. *Society and Economy* 33 (1): 29–50.
Randers, Jorgen. 2012. *2052: A Global Forecast for the Next Forty Years*. White River Junction, VT: Chelsea Green Publishing Company.
Sandén, B.A., and C. Azar. 2005. Near-Term Technology Policies for Long-Term Climate Targets—Economy Wide Versus Technology Specific Approaches. *Energy Policy* 33 (12): 1557–1576.
Sapinski, Jean Philippe. 2016. Constructing Climate Capitalism: Corporate Power and the Global Climate Policy-Planning Network. *Global Networks* 16 (1): 89–111.

Schandl, Heinz, Steve Hatfield-Dodds, Thomas Wiedmann, Arne Geschke, Yiyong Cai, James West, David Newth, Tim Baynes, Manfred Lenzen, and Anne Owen. 2015. Decoupling Global Environmental Pressure and Economic Growth: Scenarios for Energy Use, Materials Use and Carbon Emissions. *Journal of Cleaner Production* 132: 45–56.

Smith, Stephen. 2011. *Environmental Economics: A Very Short Introduction*. Oxford: Oxford University Press.

Streeck, Wolfgang. 2016. *How Will Capitalism End?* London: Verso.

Turok, Ivan, and Jacqueline Borel-Saladin. 2013. Promises and Pitfalls of the Green Economy. In *World Social Science Report 2013: Changing Global Environments*, ed. ISSC and UNESCO, 289–294. Paris: UNESCO.

UNEP. 2009. *Global Green New Deal: Policy Brief*. Geneva: UNEP.

———. 2011a. *Towards a Green Economy: Pathways to Sustainable Development and Poverty Eradication*. Geneva: UNEP.

———. 2011b. *Decoupling Natural Resource Use and Environmental Impacts from Economic Growth*. Paris: UNEP Division of Technology, Industry and Economics (DTIE).

UNFCCC. 2016. *2016 Biennial Assessment and Overview of Climate Finance Flows Report*. Bonn: UNFCCC.

Wallerstein, Immanuel, Randall Collins, Michael Mann, Georgi Derluguian, and Craig Calhoun. 2013. *Does Capitalism Have a Future?* Oxford: Oxford University Press.

WBGU. 2011. *World in Transition: A Social Contract for Sustainability*. Berlin: WBGU.

Wiseman, John, and Samuel Alexander. 2017. The Degrowth Imperative: Reducing Energy and Resource Consumption as an Essential Component in Achieving Carbon Budget Targets. In *Transitioning to a Post-Carbon Society: Degrowth, Austerity and Wellbeing*, ed. Ernest Garcia, Mercedes Martinez-Iglesias, and Peadar Kirby, 85–106. Basingstoke: Palgrave Macmillan.

World Bank. 2012. *Inclusive Green Growth: The Pathway to Sustainable Development*. Washington, DC: The World Bank.

CHAPTER 9

Identifying an Emerging Paradigm: Towards Ecosocialism?

The limitations of climate capitalism raise the issue of what alternatives could be more effective in moving society towards a decarbonised future. While capital and the vested interests that promote it continue to have an overwhelming power in structuring and driving our contemporary economy and society (Richardson et al. 2016), including how it responds to the challenges of climate change, this should not blind us to the evidence of emerging alternatives, both in practice and as proposals. For example, Erik Olin Wright identifies the major challenge today as being how to build a counter-power strong enough to curb the power of capital. Though he acknowledges Marx's 'elegant solution to this problem' that capitalism would in the long run destroy its own conditions of existence, he fails to see that today's ecological crisis shows signs of doing just that. Instead he opts for how strategies of transformation 'have long-term prospects for eroding capitalist power relations and building up socialist alternatives' (Wright 2013: 20). Among the strategies he identifies are the interstitial which builds new forms of social empowerment in the niches and margins of the dominant system, and the symbiotic which extends and deepens institutional forms of social empowerment involving both state and society simultaneously to solve practical problems. Both strategies are evident today as building pathways towards a post-carbon society in a context

P. Kirby, T. O'Mahony, *The Political Economy of the Low-Carbon Transition*, International Political Economy Series,
DOI 10.1007/978-3-319-62554-6_9

where the ecological crisis is imposing limits on the freedom of capital (such as the need to curb emissions, and reducing the exploitation and use of fossil fuels).

This chapter identifies in examples of these emerging strategies of transformation the seeds of a new paradigm, examines some of the contours of that paradigm and assesses how it might be more successful in fashioning pathways towards a post-carbon society. It begins by assembling evidence that such a new paradigm is emerging in the niches and margins of the dominant system, offering examples of what a post-carbon society might look like and what we need to do to get there. The following section outlines debates on the need to move to a degrowth economy and society, what this might entail and how feasible it is. Section three turns to one of the great problems of contemporary capitalism and an essential condition for moving to a post-carbon society, namely the structural nature of socio-economic inequality and how to reduce it. Section four maps out a way of organising an 'economy for the common good', based on different social values and governed in a much more democratic way. Taken together, these sections illustrate how the seeds of an alternative society are being sown, how they could be further developed to organise more of the economy and of society, and how they give expression to the principles of a degrowth economy. The following section discovers in the writing of the utopian socialists of the early nineteenth century many of the principles that are now motivating and finding expression in the emerging alternative. For this reason, it is being given the name of 'ecosocialism' by some analysts. The final section seeks to assess the prospects that the dominant approaches outlined in the previous chapter and the alternatives outlined in this chapter hold for moving towards a post-carbon society. A scorecard will offer an exploration of the potential future sustainability outcomes of alternative approaches to the political economy of the low-carbon transition.

An Emerging Paradigm

Reference has already been made in Chap. 4 to Belgian political scientist Olivier De Schutter emphasising the 'role of social innovations empowering people to invent local solutions' (De Schutter 2014: 17). 'These social innovations abound,' he states, 'and they are often local and territory-based, bringing together municipalities, the private sector, the "third sector" and non-governmental organisations or citizens' groups' (ibid.).

Examining the economics of sustainable consumption, Seyfang identifies 'a variety of social innovations as well as innovative technologies' proliferating in different arenas and at different scales. These take different forms 'from furniture-recycling social enterprises to organic gardening cooperatives, low-impact housing developments, farmers' markets and community composting schemes' (Seyfang 2011: 63). She distinguishes these from commercial business forms as favoured by mainstream policy, driven by interests of profit 'to appropriate the benefits of innovation in order to move ahead of the competition and so capture market rents'. This is the dominant approach as outlined in Chap. 8 which seeks 'greener' markets through taxes, incentives and better information encouraging technological innovation to improve resource efficiency and so decouple growth from environmental degradation. By contrast, grassroots innovations exist within the social as distinct from the market economy and can be seen as part of an emerging 'socio-economic system geared towards quality of life rather than economic growth per se, [which] favours localised, self-reliant economies as the basis of sustainable communities' (ibid.: 74). It finds expression in diverse organisational forms such as cooperatives, voluntary associations, mutual, informal community groups and social enterprises; meeting social and environmental needs is their primary function and an ideological commitment to alternative ways of doing things is another driver, writes Seyfang.

These niche activities therefore stand 'as a symbolic embodiment of alternatives'. This is because they represent 'the bottom-up generation of alternative systems of provision, vertical commodity chains (comprising production, marketing, distribution, retail and consumption in social and cultural context)' mediating and linking alternative forms of production with alternative and sustainable consumption (Seyfang 2011: 76). But are they fated to remain simply a marginal series of activities, often struggling to survive amid the pressures of the mainstream market-based economy, and subject to the vulnerabilities of changing public policy and funding streams? The answer to this depends on the extent to which these niche activities grow and develop so that their impact and reach greatly expands and deepens. This can happen in various ways: through scaling up as they grow in scale, through replication as they multiply in number, and through translation as mainstream activities begin to take them on. While they hold potential for spreading new ideas about sustainable production and consumption into the wider society, they also run the risk of becoming diluted in the process, coming to look more like the mainstream and losing some of their oppositional and alternative edge. Yet, as problems proliferate in the dominant socioeconomic system associated with financial crises and economic recession, cutbacks in public services due to austerity, rising oil prices and associated challenges of divesting from fossil fuels (see Box 9.1), so do the opportunities grow for grassroots initiatives to offer solutions for local communities, building resilience and empowerment. De Schutter therefore sees local innovation as offering more than niche solutions but having potentially transforming effects through promoting new economic models based on the economics of sharing, transforming how public administration, market actors and the 'third sector' understand their relationships to communities, transform social relationships and contribute to a new form of participative democracy (De Schutter 2014: 20–28).

Could it be that these transforming effects are already underway, opening spaces for a new economy and society to emerge? This is what Paul Mason calls postcapitalism based on what is happening within the current system: 'Almost unnoticed, in the niches and hollows of the market system, whole swathes of economic life are beginning to move to a different rhythm. Parallel currencies, time banks, cooperative and self-managed spaces have proliferated, barely noticed by the economics profession, and often as a direct result of the shattering of old structures after the 2008 crisis.' This has resulted in 'new forms of ownership, new forms of lending, new legal contracts: a whole business subculture has emerged over the past ten years which the media has dubbed the "sharing economy". (Mason 2015: xv). While this new society, made possible by the rise of non-market exchange and peer-production through information technologies, might in more stable times emerge slowly and progressively, it is the crisis of the present that makes more decisive action necessary, according to Mason. The financial crisis and its legacies, the starkness of climate change and the potential of technologies not realised because they are trapped within outdated organisational structures, 'mean that it will require more planning and more state ownership than anybody expects or even wants' (ibid.: 261). But he makes clear that he doesn't expect such 'revolutionary reformism' to be initiated by the state when he adds: 'So we need to inject into the environmental and social justice movements things that have for twenty-five years seemed the sole property of the right: willpower, confidence and design' (ibid.: 262).

Here we have the seeds of a new political economy model emerging. It is not the old command and control state of communism that Mason is envisioning but something more like the nervous system of a networked economy and society, creating the conditions for innovation to flourish throughout society as exemplified by Wikipedia, Open Source, open information standards and low-carbon energy installations, but within strict limits that ensure low-carbon sustainability. With this new interrelationship of state, market and society 'solutions can be found through a mixture of small-scale experiment, proven models that can be scaled up and top-down action by states' (ibid.: 267). So the 'Wiki-state' which will expand collaborative work and suppress or socialise monopolies, including the financial sector, will go hand-in-hand with the disappearance of market forces as they become redundant to allocate resources through the mechanism of supply and demand becoming instead 'the transmitter of the "zero marginal cost" effect, which manifests as falling labour time across society' (ibid.: 279). With a basic income for all, automation will take over those low-paying menial jobs that nobody wants while unleashing a wave of innovation in 'cooperative, self-managed, non-hierarchical teams' that are the most technologically advanced form of work (ibid.: 287).[1]

Not only does this apply to urban industry and services but also to agriculture. Vandana Shiva identifies a triple crisis—climate, energy and food—yet the solutions being offered such as renewable energy and various 'technological fixes' (Shiva 2016: 31) are false solutions because they derive from the same mechanistic mindset that has created the problem, backed by corporate power which benefits from them. 'Industrialization of food and agriculture has put the human species on a slippery slope of self-destruction and self-annihilation', she writes. Industrialised agriculture and the globalised food systems on which Western consumers largely depend have been promoted as the source of cheap and abundant food but instead, she argues, they are aggravating climate change through their use of chemical fertilisers and the food miles they embody, destroying local food economies, undermining food security especially for the poor, and promoting a monocultural agriculture that depletes the soil and increases our vulnerability. She states that, by contrast, 'the movement for biodiverse, ecological, and local food systems simultaneously addresses the crises of climate, energy, and food. Above all, it brings people back into agriculture and reclaims food as nourishment and the most basic source of energy' (ibid.: 143–144). Central to Shiva's claim is that 'biodiverse, organic farms and localized food systems offer us security in times of climate insecurity, while producing more food, producing better food, and creating more livelihoods' (ibid.: 109). These claims are substantiated by research. In summarising a survey of research findings on the benefits of organic agriculture, Moore states that the agri-food system 'is part of a disfunctioning planetary system that needs to be urgently addressed, particularly in the areas of nitrogen cycle, biodiversity and climate change:'

> In each of the following areas: biodiversity, soil, landscape, ground and surface water, climate/air, energy and yield, organic farming suggests solutions to be taken on by the conventional sector, or points to itself as an alternative to said conventional sector. Yield remains controversial, but only in the wealthiest parts of the world: in marginalised areas, organic is a step up from subsistence. If however we continue with a business-as-usual model of production and consumption in the agri-food system, with massive and damaging externalities, agriculture will continue to add significantly to the transgression of planetary boundaries'. (Moore 2012: 18)

Shiva's promotion of local, diverse, organic food systems produced within flourishing local economies, is consistent with Mason's vision of a postcapitalist society driven by networks of innovation protected by strong state regulation to break up or socialise the large corporations that currently dominate the global food system. And the 'transition to an age beyond oil' has to be driven by a paradigm different to the one that created the crisis, she writes (Shiva 2016: 133). So what might some of the core principles of this new paradigm be? The following sections examine degrowth, the challenge of reducing inequality, and creating an economy for the common good.

Box 9.1: Divestment: Making Moral Pressure Hit Fossil Fuel Investment

'People of conscience need to break their ties with corporations financing the injustice of climate change', said Archbishop Desmond Tutu, in response to King's College London's initial refusal to join the growing movement of UK universities divesting from fossil fuels. In response, KCL agreed to drop its most polluting investments and to join the 43 UK universities that by the end of 2016 had announced they were divesting, thus withdrawing more than £10bn from fossil fuel investments.

The campaign to divest began in the US where 35 universities had joined it by the end of 2016. The number has now been overtaken by UK universities, including leading academic institutions such as Oxford,

Edinburgh and the London School of Economics. Financial institutions and charities are also divesting and it is estimated that at least $2.6tr of assets are covered by such pledges globally (Carrington 2016a).

Overall the value of investment funds committed to selling off fossil fuel assets had reached $5.2tr by early 2017, doubling in just over a year. A report at the end of 2016 found that 688 institutions and more than 58,000 individuals across 76 countries were committed to divesting. 'Divestment has permeated every sector of society: from universities and pension funds, to philanthropic and cultural institutions, to cities, faith groups, insurance companies, and more', said May Boeve, executive director of 350.org, which has played a leading role in the campaign (Carrington 2016b).

The Challenge of Degrowth

It is paradoxical that, as concern grows within mainstream economics about 'secular stagnation' or the fear of long-term low or no growth (Streeck 2016: 65–67), degrowth is emerging to challenge the growth paradigm. The economy that emerged from the industrial revolution was premised on limitless growth. Despite some far-seeing advice from John Stuart Mill in *Principles of Political Economy* published in 1848 both that 'the increase of wealth is not boundless' and that 'the stationary state of capital and wealth' would be far preferable, with attention devoted to greater distribution and to 'moral and social progress' (Mill 1970: 111, 113, 116), belief in economic growth has become deeply embedded in modern society. Such publications as the Club of Rome's report on the limits to growth in 1972 failed to dent this almost religious belief. However, the evidence on decoupling growth from emissions surveyed in Chap. 8 raises in an ever more urgent way the necessity for recognising these limits and finding practical ways of living within them. The fundamental question has been posed by Jackson: 'How—and for how long—is continued growth possible without coming up against the ecological limits of a finite planet?' (Jackson 2009: 6).

Responding to the limits to growth debate in the 1970s, US economist Herman Daly was a pioneer of ecological economics. Daly contrasts a 'steady-state economy' with a 'growth economy'. Growth requires 'an increase in the physical state of the matter/energy throughput that sustains the economic activities of production and consumption' whereas in a steady-state economy 'the aggregate throughput is constant, though its allocation among competing uses is free to vary in response to the market'. Qualitative improvement can happen through technological innovation or through 'a deeper understanding of purpose'. Therefore a steady-state economy can develop but it 'cannot grow, just as the planet earth, of which it is a subsystem, can develop without growing' (Daly 1996: 31). For Daly, the founding assumptions of neoclassical economics no longer hold since it developed in an 'empty world' in which the economy was small in relation to the ecosystem in which it is embedded. With economic activity now breaching ecological limits, he sees the need for a new economic paradigm in which 'adjustment is by qualitative development, not quantitative growth' (ibid.: 4).[2] Daly also co-developed the Index of Sustainable Economic Welfare (ISEW) which is proposed as a more valid measure of welfare than GDP growth as the latter mainly captures quantitative growth rather than qualitative development.

While ecological economics has established itself as a subdiscipline on the margin of economics, the concept of degrowth which has more recently emerged has become a social movement (see Box 9.2). It has emerged at the intersection of the academy and social activism and some of its leading proponents profess themselves 'sceptical of the notion of the "steady-state", which focuses on the biophysical dimension and evades hard political and social questions' (Flipo and Schneider 2015: xxv). The term *décroissance* first emerged in French debates in the late 1990s and early 2000s though its origins go back to the 1970s. Originally coined as part of the limits to growth debate, it reemerged more recently as a critique of sustainable development and of growth, and is defined as 'a frame that connects diverse ideas, concepts and proposals' (Kallis et al. 2015: 4). This includes a critique of growth, of capitalism, of GDP and of commodification, but also the promotion of a reproductive economy of care, the reclaiming of the concept of the commons, support for eco-communities and cooperatives, and arguing for such policies as work-sharing and a basic income. Thus, as degrowth thinkers regularly emphasise, degrowth is not to be equated with recession in a growth economy which, as Latouche puts it, 'plunges our societies into disarray ... there is nothing worse than a growth-based society in which growth does not materialize' (Latouche 2009: 8). It quickly spread to Italy (*decrescita*) and to Spain (*decrecimiento* in Spanish and *decreixement* in Catalan). The term degrowth was first officially used in English at the Paris degrowth conference in 2008, according to Kallis et al. (ibid.: 3). They state that degrowth is 'a deliberately subversive slogan', opening possibilities to discuss alternative futures:

> Of course some sectors, such as education, medical care, or renewable energy, will need to flourish in the future, while others, such as dirty industries or the financial sector shrink. The aggregate result will be degrowth. We prefer also to use words such as 'flourishing' when we talk about health or education, rather than 'growth' or 'developing'. The desired change is qualitative, like in the flourishing of the arts. It is not quantitative, like in the growth of industrial output. (Kallis et al. ibid.: 5)

Degrowth, therefore, is about understanding the limits to growth, as is steady-state economics, but it is also about autonomy 'from the large techno-infrastructures and the centralized bureaucratic institutions, public or private, that manage them' (ibid.: 8), about repoliticisation in the sense of imagining and enacting alternative visions of a future society after capitalism (what Latouche calls 'the society of frugal abundance' (Latouche 2011)), and about how that transition can be carried out, including welfare institutions, money and credit institutions, and the politics of a degrowth transition.

Box 9.2: Degrowth as a Social Movement

Though it emerged in activist circles in France in the early 2000s, the degrowth movement is now active in more than 30 countries, mostly in Europe but also in North and South America. Its presence is most visible in France where the magazine promoting its ideas, *La Décrossance, le journal de la joie de vivre* sells 30,000 copies a month and where the Parti Pour La Décroissance (PPLD) exists as a political party.

The academic collective Research & Degrowth was also founded in France and has held a number of international academic conferences including Paris (2008), Barcelona (2010), Montreal (2011), Venice (2012), Leipzig (2014) and Budapest (2016). The Institute of Environmental Science and Technology (ICTA) at the Autonomous University of Barcelona has become a centre for degrowth research and publication, and it has helped create links with Latin American networks.

Articles on degrowth now appear regularly in mainstream academic journals and in leading newspapers including *Le Monde*, *El País*, *The Guardian*, *The Wall Street Journal* and *The Financial Times*. Courses on degrowth are taught in universities and institutes including Sciences Po in Paris and the School for Oriental and African Studies (SOAS) in London.

However, proponents of degrowth acknowledge the continuing need for economic growth in countries of the South if the majority of their population is to achieve an improved standard of living. Thus degrowth in the North 'will liberate ecological space for growth in the South' state Kallis et al. (2015: 5). This is echoed by Pope Francis who states in the encyclical letter *Laudato Sí* that 'the time has come to accept decreased growth in some parts of the world, in order to provide resources for other places to experience healthy growth' (Pope Francis 2015: par. 193). To help decide which countries can grow while others degrow, Dietz and O'Neill provide a quadrant on two axes—the horizontal measures resource use while the vertical measures the economy's size. This provides four grids:

- Undesirable growth: large economies with growing resource use;
- Desirable degrowth: large economies with decreasing resource use;
- Undesirable degrowth: small economies with decreasing resource use;
- Desirable growth: small economies with growing resource use.

Making use of this in practical politics will require rigorous and reliable indicators of both economic scale and resource use over time, they state, but also defining the optimal scale of an economy, namely its 'maximum sustainable size' (Dietz and O'Neill 2013: 185–186).

Growing awareness that economic growth in its current form is not compatible with living within the limits of the ecosystem overlaps with another critique of growth emerging from a distinct set of concerns, namely that economic growth is incompatible with the good life. In their study of money and the good life, Skidelsky and Skidelsky conclude that the material conditions for the good life already exist, at least in the affluent parts of the world, but that 'the blind pursuit of growth puts it continually out of reach' since it turns wealth into an end to be pursued rather than a means to the good life which they describe as 'health, respect, friendship, leisure and so on'. Therefore 'economic growth should be accepted as a residual, not something to be aimed at' (Skidelsky and Skidelsky 2013: 13, 14).[3] The recognition that growth may undermine as well as enhance wellbeing opens the possibility that reducing consumption and production may enhance social as well as ecological wellbeing. As Alexander puts it, a degrowth transition will involve examining what is truly necessary to live a dignified life 'as well as letting go of so much of what is superfluous and wasteful in consumer society today'

(Alexander 2017: 158). This dimension of the lived reality of degrowth has been neglected, he argues:

> A degrowth economy may be 'austere' (but sufficient) in a material sense, especially in comparison to the cultures of consumption prevalent in developed regions of the world today. But such austerity could also liberate those developed or over-developed societies from the shackles of consumerist cultures, freeing them from materialistic conceptions of the good life and opening up space for seeking prosperity in various non-materialistic forms of satisfaction and meaning (ibid.: 158).

Degrowth has been useful in asking deeper questions about the ends of economic life and about distinguishing human needs from human wants, issues that touch on values and draw philosophical and indeed religious contributions into the frame. And, as one of the elements recognised by Skidelsky and Skidelsky as damaging 'the moral fabric of society' is inequality, both material inequality and its effects on the equality of respect that underpins our democratic life, to this we now turn (Skidelsky and Skidelsky 2013: 159–160).

The Challenge of Equality

As De Schutter emphasises, 'equality is not simply an end which is valuable in itself, as a component of social justice ... it is also a means to accelerate the transition to more sustainable societies' (De Schutter 2014: 16). He offers three reasons for this positive relationship between equality and sustainability:

- Inequality fosters status competition in which the desire to emulate those who have more fuels consumption;
- The pursuit of greater equality would mean putting a brake on the most unsustainable lifestyles, especially those of the most affluent;
- Support for policies to reduce emissions depends on them being seen as fair, as applying equally and effectively to those sectors most responsible for them.

'Equal societies are better equipped to transform themselves: the more a society is equal, the fewer and the least powerful are the groups who will have strong reasons to oppose change', concludes De Schutter (ibid.: 18).

While the reduction of economic inequality became a major objective of social policy in Western countries in the 1960s and early 1970s through progressive taxation and redistribution measures, this was replaced in the late 1970s and 1980s by a focus on equality of opportunity, particularly on the basis of gender and race. At the same time, economic inequality began to rise in the US and other Western countries after a period that had seen it sharply decline. This rise came to be the focus of academic attention in the 1990s and, since the financial crisis of 2008–2009 'the rise in inequality has become a major political issue' (Galbraith 2016: 4), reflected in a slew of high-profile books (Wilkinson and Pickett 2009; Stiglitz 2013; Piketty 2014; Atkinson 2015). With the UK vote on Brexit and the election of Donald Trump in 2016, rising inequality came to be seen as a contributory factor requiring decisive action (Box 9.3). Highlighting the gravity of the situation, on the eve of the 2017 World Economic Forum in Davos, Oxfam published a briefing paper containing evidence that eight men now own the same amount of wealth as the poorest half of the world and that over the next 20 years, 500 people will hand over $1.2 trillion to their heirs, a sum larger than the GDP of India (Oxfam 2017).

To be effective, action on growing inequality depends on understanding its causes. Recent analysis treats it as a multidimensional issue, deeply embedded in structural features of contemporary society. Atkinson lists six contributing factors: globalisation, technological change, changing pay norms, the reduced role of trade unions, the growth of financial services and the scaling back of redistributive tax and transfer policies (Atkinson 2015: 82). The issue of technological change in a globalised economy relates to trends in wages for skilled and unskilled workers and the ability to outsource production to more low-wage economies. In this situation, pay norms change as technology replaces unskilled workers and as those with higher skills can bargain for higher wages, thereby increasing inequality. Globalisation and economic liberalisation have also greatly weakened the bargaining power of trade unions and resulted in a marked decline in the unionisation of the workforce throughout the world. The growth of financial services reflects the liberalisation of the financial sector, the integration of financial markets globally and the growth in the power of banks and other financial agencies. Galbraith regards this as having 'played a powerful role affecting economic inequality around the world' (Galbraith 2016: 111) for two main reasons: the increase in incomes in the financial sector, and concentrating the growth of investment and associated income among a small number of players. Finally, there has been a tendency in most developed countries towards reducing rates of taxation on higher earners and on capital, and towards curbing welfare benefits, often linking them to labour market activation measures. Yet, though deeply embedded in the way contemporary societies are structured, Atkinson cautions against 'creating the impression that inequality is rising on account of forces outside our control'. He adds:

> It is my belief that the rise in inequality can in many cases be traced directly or indirectly to changes in the balance of power. If that is correct, then measures to reduce inequality can be successful only if countervailing power is brought to bear. (Atkinson 2015: 82–83)

Most proposals to reduce inequality begin with policy measures centred on reducing the inequality of market incomes through measures such as minimum wages, combined with more effective redistribution measures, such as tax and transfer programmes (see Galbraith 2016: 135–136 and Atkinson 2015: 110). A further element, mentioned by Galbraith, refers to changing the cost of living through taxing the sale of commodities or providing low-cost public goods. Such measures, however, depend on a fundamental change in the relative power of the state and of market actors, particularly as they have developed in the era of neoliberalism. Atkinson speaks of the need 'to identify the locus of decision-making as it affects the incomes and lives of individuals, as well as the balance of power—between individuals and between groups in society'. Re-constituting the role of the state is therefore essential to addressing inequality, not only through its role in redistributive taxation and the provision of social security, but also through the influence it can exercise on market incomes (Atkinson 2015: 110–111). Atkinson's 15 proposals to 'substantially reduce the extent of inequality' go far beyond the usual focus on tax and transfer measures to include the direction of technological change through public investment policies which are designed to enhance employability rather than undermine it, and measures to transfer power in the direction of consumers and restore the legal position of trade unions thus strengthening powers countervailing that of capital. Other proposals to strengthen the role of the state include guaranteed public employment at the minimum wage, a national pay policy that consists of a statutory minimum wage and a code of practice for pay above the minimum, a guaranteed positive real rate of interest on savings via national savings bonds, a capital endowment paid to all at adulthood, a public Investment Authority 'operating a sovereign wealth fund with the aim of building up the net worth of the state by holding investments in companies and in property' and a range of measures on progressive taxation (marginal rates increasing to a top rate of 65%), on inheritance and gifts and on property tax. Furthermore, he proposes a Child Benefit 'at a substantial rate', a 'participation income' (a guaranteed basic income but conditional on the recipient making a social contribution to underpin an ethic of reciprocity) and a renewal of social insurance. Finally, to address global inequalities, he proposes that the target for Official Development Assistance should be raised to 1% of Gross National Income (it is currently 0.7% of GDP, a target so far reached by only five countries) (ibid.: 237–239). Taken together, these measures would not only address inequality but would be consistent with the emerging alternative paradigm surveyed so far in this chapter.

Box 9.3: Piketty and Hawking on Inequality, Climate Change and Trump

'The main challenges of our times are the rise in inequality and global warming', wrote French academic and writer Thomas Piketty following the electoral victory of Donald Trump. This was primarily due to 'the explosion in economic and geographic inequality' in the US over several decades and the inability of successive governments to deal with it (Piketty 2016). Physicist Stephen Hawking also sees the Brexit vote in the UK and the election of Trump as 'a cry of anger' by people who feel abandoned and who are suffering under the socially destructive nature of widening economic inequality around the world.

The lesson is clear, 'as a matter of urgency, globalisation must be fundamentally re-oriented', writes Piketty. Yet, both the Clinton and Obama administrations not only went along with the market liberalisation introduced by the Reagan and both Bush presidencies, but 'at times they even outdid them'. International treaties are now necessary to respond to these challenges and 'promote a model for fair and sustainable development'. 'With resources increasingly concentrated in the hands of a few, we are going to have to learn to share far more than at present', writes Hawking (Hawking 2016).

Trump's programme will only 'strengthen the trend towards inequality' if he abolishes the health insurance granted to low-paid workers and reduces corporation tax from 35% to 15%, setting the US 'on a headlong course into fiscal dumping'. Instead, what is necessary are public services and infrastructure, health and education systems, and fair taxation systems. 'If we fail to deliver these, Trumpism will prevail' (Piketty 2016). Hawking see this as 'the most dangerous moment in the development of humanity' as we have the technology to destroy the planet.

Economy for the Common Good

In discussing the feasibility and affordability of his proposals to address inequality, Atkinson ends with an emphasis on the power of citizens, their values such as fairness and social justice, and on the need to initiate a 'national conversation' about national goals, such as setting a target for unemployment (Atkinson 2015: 305–308). So, looking beyond the important challenge of equality, what sort of economy might citizens opt for and what values would inform their options? Felber identifies 'an ethical schizophrenia created by the chasm between business and society' (Felber 2015: xviii), the former operating on values of egoism, greed, avarice, envy, ruthlessness and irresponsibility that contradict the values of trust, honesty, esteem, respect, empathy, cooperation, mutual help and sharing that most people report as aspiring to guide their daily interpersonal relationships. He concludes therefore that our current market economy needs to be put on a new course: 'directing our path away from pursuit of profit and competition, and instead striving towards pursuit of the common good and cooperation' (ibid.: 17). This he entitles the Economy for the Common Good but he emphasises that this strives to be combined with other alternatives such as the solidarity-based economy, the commons, economic democracy and the post-growth economy, and he explicitly mentions alternatives developed by Tim Jackson, the Skidelskys and Herman Daly. Felber's outline of an economy for the common good therefore integrates into a comprehensive model of an alternative economy many of the elements already identified in this chapter.

In Felber's outline, the common good becomes the goal of the economy and the criterion for its success. 'Neither the use of money nor the increase thereof would be compulsory—the success of enterprise, investment and national economies would not be measured in terms of profit but rather in terms of the goal of promoting the common good' (ibid.: 18). This has five essential components: human dignity; cooperation and solidarity; ecological sustainability; social justice; and democratic co-determination and transparency. These are combined into a Common Good Balance Sheet that provides ways of measuring success on each of these components at three levels: investment, the enterprise and the national economy (see Felber 2015: 26–27 for a matrix). This should have binding force on all enterprises, be holistic in that it applies to all aspects of business, be measurable allowing objective evaluation, be comparable across all enterprises, be comprehensible to all citizens, be easily available, be subject to external audit and, finally, have legal consequences. These eight requirements 'could have the desired effect of ethically rerouting the economy in the direction of sustainability, distributional justice and meaningful, health-promoting labour' (ibid.: 30). Not only would this make full information on all companies' ethical practices available to consumers through an easy-to-use colour-coded labelling system, but it would enable legal privileges such as a lower tax rate or better loan conditions to be offered to the most ethical, thereby allowing ethical and responsible enterprises gain a stronger foothold in the market. 'The "laws of the market" would be harmonized with the basic values of society', writes Felber (ibid.: 33).

This economy would require an entirely different financial system in which money as credit would become a public good and financial markets would be closed. A Democratic Bank, under citizens' control, would assume the core functions of the financial markets including personal accounts, inexpensive loans for enterprises and private households as well as the creation of ecological and social added value through investments for the common good, a full-scale network of branches, and inexpensive supplementary loans to the state. The bank would neither charge nor pay interest and would finance itself through lending fees to cover its core costs. Since the only stock market would be to provide opportunities for investments with a social or ecological return, savers would have an incentive to channel deposits to the Democratic Bank so as to benefit society. Other banks would only exist in a legal form that excluded profit-making, such as cooperatives and savings banks. Felber writes that, with these reforms, 'money would be forced back into its serving role': 'No one could become rich through possession of money alone; income would be made by working and for this reason such earned income would be enough to lead a good life' (ibid.: 76).

Felber would organise the productive economy on similar lines, so as to ensure that all people and market participants could enjoy the same liberties, rights and opportunities. To avoid too much economic power being concentrated in too few hands as happens in today's economy, a regional economic parliament acting on behalf of citizens would nominate representatives to sit on the supervisory board of all large companies in the region. This would ensure that all private companies would remain small (the size would be determined democratically) since all larger companies would be subject to public supervision. As companies increase in size, so would the requirement for profits to be shared with employees. Apart from private and public enterprises, a third form of enterprise would be introduced to run publicly owned enterprises providing public services such as the railways, postal services, universities, utility companies, kindergartens and even banks by an executive board elected through direct democracy. This would be a 'democratic commons', modelled on the 'commons' that existed prior to capitalism which made assets such as forests and meadows available to all local inhabitants. Felber also proposes a number of measures to restrict individuals building up levels of income or wealth that undermine the common good. Central to these would be establishing through a democratic economic convention the limits for income inequality: this would establish the relationship of maximum to minimum incomes such as 7, 10, 12 or 20-fold. Any increase in the maximum income would therefore require an increase in the minimum income to maintain the ratio. Other proposals include the democratisation of corporations and profit sharing with employees, restricting the right to inherit to a level democratically decided and lodging all assets exceeding this amount in a public Generations Fund to be distributed to all members of the next generation in equal parts in the form of a 'democratic dowry'. Based on estimates for Germany, he calculates that this would amount to around €200,000 for each individual. Those receiving inheritances would have this amount deducted from their dowry. Similarly the amount of company shares that could be inherited in a family enterprise would be limited. Neither could land or any part of nature be privately owned: those requiring it for cultivation would be allocated a limited amount at no charge by the local municipality. In these ways, the unequal starting conditions and therefore power relations that characterise capitalist societies would be avoided.

An economy organised along the lines outlined by Felber would underpin and give structural coherence to the overlapping features of the new paradigm as exemplified by the various analyses/initiatives discussed in this chapter: local, small-scale, cooperative, innovative production within strong state regulation to curb the power of big corporations and of speculative capital and to create the conditions that ensure far greater equality of condition. Some of the details may be different: for example, Felber does not recommend a basic income but his proposals to establish a ratio between maximum and minimum incomes and his 'democratic dowry' would achieve the same result. On the role of economic growth, he writes that 'systematic drivers of growth should be removed from the economic order' (ibid.: 68) so as to seek to live within the carrying capacity of the Earth. He thus foresees 'permanent growth of the Common Good—health, education, co-determination, quality of the environment and relationships, safety, stability, peace—but not necessarily growth of money, and certainly not growth of natural and material resource consumption' (ibid.: 210). This is consistent with the proponents of degrowth. He doesn't outline the move from a linear to a circular economy (see Box 9.4), apart from a mention of 'notions like circulation economy' (ibid.: 148) but his proposals imply such a move. Furthermore, he offers examples of cooperative enterprises, ethical banks, agricultural projects, open-source technology and non-profit organisations in various regions of the world that already operate according to the principles of the Economy for the Common Good (ibid.: Chap. 7). The new paradigm is therefore already emerging, based on principles alternative to those that drive the mainstream dominant economy and society. Yet, its potential is often underestimated as it is seen as a series of small-scale, fragmented alternatives that carry little weight against the power of the dominant capitalist model. Does it constitute an emerging alternative in the way that capitalism emerged within the structure of feudalism, and if it does, what essentially differentiates it?

Box 9.4: Moving the EU Towards a Circular Economy

'Switching from a linear (take-make-use-throw away) economy to an eco-design focused circular (make-use-reuse-remanufacture-repair) economy in which nothing is wasted is a critical challenge for Europe', states the European Economic and Social Committee (EESC) in a June 2016 Position Paper. Acknowledging the European Commission's moves in this direction and its 2015 action plan for the circular economy, the EESC notes 'a distinct lack of ambition' in certain areas.

The circular economy needs to be 'long-lasting, small, local and clean', it states and urges the EU Eco-Design directive 'to take the full lifecycle of the product into account, including its durability, reusability, reparability, recyclability as well as availability and affordability of spare parts'. It also urges 'a total ban on products with designed obsolescence or built-in defects'. Behaviour change will require 'a shift away from the traditional concept of product ownership' as leasing and selling goods as services become standard practice (EESC 2016).

Postcapitalism: From Utopian Socialism to Ecosocialism

In themselves, most of the initiatives outlined here fit within capitalism, even if they exist on the margins of an economy dominated by large enterprises organised to maximise profit as a core principle of survival. However, they are based on principles that essentially are alternative to the principles underlying capitalism so that, if by legislative and regulatory frameworks these principles were more robustly protected and those motivating large corporate enterprises more severely constrained, these many grassroots initiatives could be the seeds of a new type of economy and society. This would continue to be a form of market economy because private enterprises, money and market-generated produce and prices would exist, but in a fundamentally different way, socially cooperative rather than profit maximising. Neither would profits cease to be important but the essential difference is that they would be used to add social and ecological value through investments in cooperative enterprises and social provision, rather than being used for personal enrichment, exploitation of workers, environmental destruction or competitive takeovers. Profits would become a means to economic and social development for the common good rather than an end in themselves for distribution to shareholders. Such an economy would contradict and make impossible Marx's defining feature of capitalism, namely that the owners of the means of production expropriate most of the value of what workers produce. As Felber writes, to move beyond this, power and responsibility would be decoupled and capital would be limited to being a means and never an end (for practical proposals, see Felber 2015: 41–43). The essential paradigm underlying the development model would shift from competition to cooperation. Felber's proposals would also contradict Karl Polanyi's definition of capitalism as a market society created through the commodification of land, labour and money. In an economy for the common good, each of these would be protected against the inroads of the market economy and thereby decommodified. As Polanyi's daughter, Kari Polanyi Levitt put it, echoing the principles of her father's work:

> If we cannot set limits to the reach of the market, economic forces will destroy the capacity of society to resist disintegration and the capacity of the biosphere to renew itself. Public ownership and social and economic planning must be rescued from their current status as heresies. The vision of socialism as a co-operative, democratic and just economic order based on the social ownership and control of natural and man-made resources, united by the enjoyment of a community of culture, embodies the best of the legacy of the European enlightenment. (Polanyi Levitt 2013: 53)

Naming this new form of society as a form of socialism establishes its distinctive nature. However, it is a different form of socialism to that which dominated throughout the twentieth century, namely a statist and often authoritarian and repressive form, highly destructive of the ecosystem. This is recognised in Alex Honneth's attempt to reformulate the original intention of socialism 'so as to make it once again a source of political-ethical orientations' (Honneth 2017: 5). Instead, as Mason recognises, it takes us back to an earlier form of what is often called utopian socialism which was sidelined by the emergence of the factory system:

> The utopian socialist communities of the mid-nineteenth century failed because the economy, technology and the levels of human capital were not sufficiently developed. With info-tech, large parts of the utopian socialist project become possible: from cooperatives, to communes, to outbreaks of liberated behavior that redefine human freedom. (Mason 2015: xvi)

What today are now known as utopian socialists were a group of somewhat disparate thinkers and organisers in the early part of the nineteenth century including, in France Claude Henri de Saint-Simon (1760–1825), Charles Fourier (1772–1837) and Étienne Cabet (1788–1856), in Britain Robert Owen (1771–1858), and in Germany Wilhelm Weitling (1808–1871). They never saw themselves as utopians; this was a label given dismissively to them later by Karl Marx and Friedrich Engels, to distinguish them from the scientific socialism being developed by these latter thinkers. However, despite the many differences among the earlier socialists, common themes and issues characterise their writings and work which have a resonance today. Taylor identifies a common goal, namely the achievement of a harmonious society; many of them created communities with the title Harmony such as Owen's experimental communities New Harmony in Scotland and the US. This was to be achieved through three key means: association, namely creating better working conditions through developing bargaining power and even communal property, particularly among workers; community, often through establishing actual communities in distinct locations or through friendly societies or trade clubs; and cooperation, a principle seen as offering an alternative to capitalism (Taylor 1982: 3–9). He lists what he calls 'six strategic dilemmas' that run through their thinking: (i) industrialism versus anti-industrialism; (ii) private property versus common ownership; (iii) religion versus secularisation; (iv) revolution versus gradualism; (v) statism versus communitarianism; and (vi) democratic versus authoritarian organisation (ibid.: 9–18). Despite these differences, what distinguishes them from the legacy of Marx and Engels is a distinctive view of social revolution. While the socialist tradition that dominated the twentieth century saw social revolution as happening through the seizure of state power through a vanguard party, the earlier utopian socialists saw it much more as, in the words of Toivanen, 'the production of social power that creates common value practices and forms of life that dissolve the old society … in a way that "cracks" also the established political institutions and practices that try to discipline, control and command the existing society in the interest of capital' (Toivanen 2015: 128). This form of socialist practice being promoted as a response to the challenges of the socioecological transition to a post-carbon society, is increasingly being labelled ecosocialism (for a brief history see Gonick 2010).

Kovel defines ecosocialism as 'that society in which production is carried out by freely associated labour and with consciously ecocentric means and ends'. When such production takes hold across society as a whole, it can be recognised as a mode of production so that 'ecosocialism will be a society whose mode of production is ecocentric'. What Kovel calls 'the coordinated agencies of society' such as the state, civil society, culture and religion would all be 'centered about ecocentric production' which 'hems in markets and keeps them functioning according to ecocentric ethics rather than profiteering'. 'Use-value and quality are valorized over exchange-value and quantity, and the economy is now embedded within society rather than, as under capitalism, standing over society' (Kovel 2007: 243). This definition employs some key concepts of the Marxist tradition while integrating these into a framework drawn from political ecology. According to Löwy, it rests on two essential arguments: firstly, that the present mode of capitalist production and consumption is based on a logic of boundless accumulation that is resulting in the accelerated destruction of the environment, and secondly, that the expansion of this civilisation based on a market economy threatens the very survival of the human species (Löwy 2005: 18–19). However, the integration of ecological concerns into a Marxist framework continues to excite debate so that, in the words of Burkett 'we still do not have anything approaching a deep conceptual synthesis of green and red theories, one based on a vital awareness of the historical development of each' (Burkett 2006: 23). More fruitful though less actively worked have been attempts to marry ecological concerns with Polanyi's conceptual framework (Adaman et al. 2007; Kirby 2013: 56–61). Despite the continuing conceptual debates, ecosocialism is emerging both as a recognised political stance (see Box 9.5) and as a way of conceiving of alternative futures (Santiago Muíño 2016; Taibo 2016; Löwy 2011; Kovel 2007).

Box 9.5: Ecosocialism as a Political Option in Spain

Joan Ribó of the left-wing Compromis party, a month after becoming mayor of the Spanish city of Valencia following 24 years of rule by the conservative Partido Popular (PP), was asked if he was a nationalist. He answered: 'I am not a nationalist though I defend the national liberties of my country. If I am anything, I am an ecosocialist' (Serra 2015).

This reflects the growing current of ecosocialism within Spanish left-wing politics. The 3rd International Ecosocialist Conference held in Bilbao in September 2016 organised by 16 trade unions, NGOs and political parties including Podemos, pledged to lead 'an eco-social transition process'. This requires 'a radical democratic change in certain means of production and consumption which puts in the central position of life the people's basic needs, which should be determined democratically and in accordance to the biophysical limits of the planet (ecosocialism)'. (Third International Ecosocialist Conference 2016).

Towards a Low-Carbon Transition, Assessing the Prospects of the Political Economy Approaches

Scenarios are a useful tool to analyse plausible future changes in global dynamics, political themes and the impacts of these future worlds on domains of interest. They allow us to 'think out of the box' and view the future as a realm of possibility, to explore alternative pathways and what they may mean for sustainability concerns of society, environment and economy. The Tellus Institute scenarios (Raskin et al. 2010), referred to in Chap. 5, were developed as a set of global scenarios for the century ahead that explore alternative sustainability outcomes. They provide a suitable basis to consider 'climate capitalism' and 'ecosocialism' approaches to the low-carbon transition and the political economy of the future. Climate capitalism could be linked to the Market Forces (MF) 'conventional world' of the future dominated by technological responses to environmental challenges and a free market approach to implementation. The 'Great Transition' (GT) scenario is an 'alternative vision' that could be used to explore an ecosocialist political economy, where social and environmental wellbeing is prioritised through a sustainable development approach, initiated both bottom-up through society and top-down through government intervention. A set of three assessment criteria for each of the key domains for sustainable development is used: environmental, social and economic. Environmental outcomes include GHG emissions reduction, climate change impacts and ecosystem protection. Social outcomes include individual human wellbeing, social equality and the strength of society and community. Economic outcomes explored include economic growth, the economic costs of low-carbon transition and the extent of technology deployment. Each of these nine domains is scored on a five-point scale for the degree of impact of the approaches detailed in Table 9.1.

The outcomes are consistent with global scenario literature and reviews of the driving forces of development including transition literature such as the World Bank's report (2015) on the ability of the technological transition to effectively decarbonise the global economy, as reviewed in Chap. 5. They are also consistent with global scenario literature that determines that emissions' reductions have a stronger association with higher government intervention (Morita and Robinson 2001: 141), and that more desirable outcomes in general are associated with sustainable development (Sachs 2016). In contrast to Raskin et al. (2010) it is assumed here that both of these political economy approaches successfully reduce emissions in keeping

Table 9.1 Sustainability outcomes of alternative approaches to the political economy of the low-carbon transition

Scenario	*'Market forces'*	*'Great transition'*
Political economy	Climate capitalism	Ecosocialism
GHG emissions reduction	● ● ● ●	● ● ● ● ●
Climate change impacts	● ● ● ●	● ● ●
Ecosystem protection	● ●	● ● ● ●
Individual human wellbeing	● ●	● ● ● ● ●
Social equality	●	● ● ● ● ●
Strength of community and society	●	● ● ● ● ●
Economic growth	● ● ● ●	● ●
Economic cost of low-carbon transition	● ● ● ● ●	● ●
Extent of technology deployment	● ● ● ● ●	● ●

with the +2 °C target under the UNFCCC, although a 'development approach' employed through ecosocialism more deeply reduces GHG emissions at a fundamental level and does not just rely on technology. With ecosocialism the benefits are higher and costs lower, leading to less climate change and better social and environmental outcomes. However, it is not only the processes of implementation that differentiate these scenarios but the actual development outcomes. Ecosocialism would tend to score higher on the desirable environmental and social outcomes than climate capitalism. Climate capitalism illustrates stronger economic growth and technology deployment, but the costs of low-carbon transition would be higher and the benefits of the growth distributed unevenly. This is an indicative illustration of the potential outcomes from the two political economy approaches detailed, but one which leaves room for thought on our approaches to the low carbon transition on which we reflect in Chap. 10.

Notes

1. While noting that this could also be regarded as a form of techno-optimism, it is not guaranteed that such automation, robotics and proliferating technological advancement will resolve the politics of distribution, the environmental impacts of production and consumption or the vicissitudes of the system in general.
2. Growth can occur in value rather than physical throughput and qualitative growth is also a possibility. This could occur where economic growth arises in human capital and the preservation and enhancement of natural capital. The potential of such developments is unknown with respect to establishing limits to growth. Jakob and Edenhofer (2015) suggest that focussing on welfare could render the growth versus degrowth debate defunct.
3. Growth as means and not the ends of development is a position articulated in the capability approach to human development (Anand and Sen 2000).

References

Adaman, Fikret, Pat Devine, and Begum Ozkaynak. 2007. Reinstituting the Economic Process: (Re)embedding the Economy in Society and Nature. In *Karl Polanyi: New Perspectives on the Place of the Economy in Society*, ed. Mark Harvey, Ronnie Ramlogan, and Sally Randles, 93–112. Manchester: Manchester University Press.

Alexander, Samuel. 2017. Frugal Abundance in an Age of Limits: Envisioning a Degrowth Economy. In *Transitioning to a Post-Carbon Society: Degrowth, Austerity and Wellbeing*, ed. Ernest Garcia, Mercedes Martinez-Iglesias, and Peadar Kirby, 157–177. Basingstoke: Palgrave Macmillan.

Anand, S., and A. Sen. 2000. Human Development and Economic Sustainability. *World Development* 28 (12): 2029–2049.

Atkinson, Anthony B. 2015. *Inequality: What can be Done?* Cambridge, MA: Harvard University Press.

Burkett, Paul. 2006. Two Stages of Ecosocialism? Implications of Some Neglected Analyses of Ecological Conflict and Crisis. *International Journal of Political Economy* 35 (3): 23–45.

Carrington, Damian. 2016a. Fossil Fuel Divestment Soars in UK Universities. *The Guardian*, 22 November 2016.

———. 2016b. Fossil Fuel Divestment Funds Double to $5tr in a Year. *The Guardian*, 12 December 2016.

Daly, Herman E. 1996. *Beyond Growth: The Economics of Sustainable Development.* Boston: Beacon Press.

De Schutter, Olivier. 2014. *The EU's Fifth Project: Transitional Governance in the Service of Sustainable Societies.* Accessed 19 February 2015. http://www.srfood.org/images/stories/pdf/otherdocuments/Framing4.pdf

Dietz, Rob, and Dan O'Neill. 2013. *Enough is Enough: Building a Sustainable Economy in a World of Finite Resources.* London: Routledge.

EESC. 2016. The Circular Economy Package. *Position Paper*, June 2016. Brussels: European Economic and Social Committee.

Felber, Christian. 2015. *Change Everything: Creating an Economy for the Common Good.* London: Zed Books.

Flipo, Fabrice, and François Schneider. 2015. Foreword. In *Degrowth: A Vocabulary for a New Era*, ed. Giacomo D'Alisa, Federico Demaria, and Giorgos Kallis, xxiii–xxxvi. London: Routledge.

Galbraith, James K. 2016. *Inequality: What Everyone needs to Know.* Oxford: Oxford University Press.

Gonick, Cy. 2010. Exploring Ecosocialism as a System of Thought. *Canadian Dimension* 44 (5): 36–42.

Hawking, Stephen. 2016. This is the Most Dangerous Time for Our Planet. *The Guardian*, 1 December 2016.

Honneth, Alex. 2017. *The Idea of Socialism.* Cambridge: Polity.

Jackson, Tim. 2009. *Prosperity without Growth: Economics for a Finite Planet.* London: Earthscan.

Jakob, M., and O. Edenhofer. 2015. Welfare with or without Growth? Do We Need to Reduce Economic Activity to Protect the Environment and Increase the Quality of Life? *GAIA* 24 (4): 240–242.

Kallis, Giorgos, Federico Demaria, and Giacomo D'Alisa. 2015. Introduction: Degrowth. In *Degrowth: A Vocabulary for a New Era*, ed. Giacomo D'Alisa, Federico Demaria, and Giorgos Kallis, 1–17. London: Routledge.

Kirby, Peadar. 2013. Towards a Post-Carbon Society: Climate Capitalism or Ecological Socialism? *Arxius* 29: 53–66.

Kovel, Joel. 2007. *The Enemy of Nature: The End of Capitalism or the End of the World?* London: Zed Books.
Latouche, Serge. 2009. *Farewell to Growth.* Cambridge: Polity.
———. 2011. *Vers une société d'abondance frugale.* Paris: Fayard.
Löwy, Michael. 2005. What is Ecosocialism? *Capitalism Nature Socialism* 16 (2): 15–24.
———. 2011. *Ecosocialismo: La alternativa radical a la catastrophe ecológica capitalista.* Buenos Aires: El Colectivo and Herramienta.
Mason, Paul. 2015. *PostCapitalism: A Guide to Our Future.* London: Allen Lane.
Mill, John Stuart. 1970. *Principles of Political Economy.* London: Penguin. [original edition: 1848].
Moore, Oliver. 2012. Arguing a Case for Organic Farming, with an Emphasis on Biodiversity. *Unpublished Paper.* Cork: Centre for Co-operative Studies, UCC.
Morita, T., and J. Robinson. 2001. Greenhouse Gas Emission Mitigation Scenarios and Implications. In *Climate Change 2001: Mitigation*, ed. B. Metz, O. Davidson, R. Swart, and J. Pan. Cambridge: Cambridge University Press.
Oxfam. 2017. An Economy for the 99%. *Oxfam Briefing Paper.* Oxford: Oxfam UK for Oxfam International.
Piketty, Thomas. 2014. *Capital in the Twenty-First Century.* Cambridge, MA: Harvard University Press.
———. 2016. We Must Rethink Globalization, or Trumpism will Prevail. *The Guardian*, 16 November 2016.
Polanyi Levitt, Kari. 2013. *From the Great Transformation to the Great Financialization.* London: Zed Books.
Pope Francis. 2015. *Laudato Si: On Care for our Common Home.* Vatican City: Vatican Press.
Raskin, P., C. Electris, and R.A. Rosen. 2010. The Century Ahead: Searching for Sustainability. *Sustainability* 2: 2626–2651. doi:10.3390/su2082626.
Richardson, H.S., Erik Schokkaert, Stefano Bartolini, Geoffrey Brennan, Paula Casal, Matthew Clayton, Rahel Jaeggi, et al. 2016. Social Progress: A Compass. *International Report on Social Progress.* Chapter 2 Draft for Comment. Accessed 5 January 2017. https://comment.ipsp.org/chapter/chapter-2-social-progress-compass
Sachs, J. 2016. Happiness and Sustainable Development: Concepts and Evidence. In *World Happiness Report 2016, Update*, ed. J. Helliwell, R. Layard, and J. Sachs, vol. I. New York: Sustainable Development Solutions Network.
Santiago Muíño, Emilio. 2016. *Rutas sin mapa: Horizontes de transición ecosocial.* Madrid: Catarata.
Serra, Maria Josep. 2015. Ribó: "Compromís y Podemos puede ser el primer partido de España". *El País*, 19 July 2015.
Seyfang, Gill. 2011. *The New Economics of Sustainable Consumption: Seeds of Change.* Basingstoke: Palgrave Macmillan.

Shiva, Vandana. 2016. *Soil, Not Oil: Climate Change, Peak Oil and Food Insecurity*. London: Zed Books.

Skidelsky, Robert, and Edward Skidelsky. 2013. *How Much is Enough? Money and the Good Life*. London: Penguin.

Stiglitz, Joseph. 2013. *The Price of Inequality*. London: Penguin.

Streeck, Wolfgang. 2016. *How Will Capitalism End? Essays on a Failing System*. London: Verso.

Taibo, Carlos. 2016. *Colapso: Capitalismo terminal, transición ecosocial, ecofascismo*. Madrid: Catarata.

Taylor, Keith. 1982. *The Political Ideas of the Utopian Socialists*. London: Frank Cass.

Third International Ecosocialist Conference. 2016. Final Manifesto. Accessed 10 January 2017. http://alterecosoc.org/wp-content/uploads/2016/10/manifiesto-inglesez.pdf

Toivanen, Tero. 2015. Commons Against Capitalism. In *The Politics of Ecosocialism: Transforming Welfare*, ed. Kajsa Borgnäs, Teppo Eskelinen, Johanna Perkiö, and Rikard Warlenius, 116–134. London: Routledge.

Wilkinson, Richard, and Kate Pickett. 2009. *The Spirit Level: Why More Equal Societies Almost Always Do Better*. London: Allen Lane.

World Bank. 2015. *Decarbonizing Development: Three Steps to a Zero-Carbon Future*. Washington, DC: World Bank.

Wright, Erik Olin. 2013. Transforming Capitalism through Real Utopias. *American Sociological Review* 78 (1): 1–25.

CHAPTER 10

Options and Prospects for a Global Low-Carbon Transition

The move to a low-or post-carbon society is usually described as a transition. Box 1.3 discussed how some analysts draw on the work of Karl Polanyi and liken it in scale to the Neolithic or the Industrial revolutions. So, as discussed in Chap. 5 and consistent with the title of Polanyi's grand opus *The Great Transformation*, what is required for human societies to transition to a state in which they can flourish within the capacity of the planet's ecosystem is a transformation that is accurately described as revolutionary. For the first time in human history, humanity faces a stark option: revolutionary transformation or potential collapse and social disintegration. This chapter explores prospects for choosing more benign pathways over the latter.

It begins by summarising the argument of the book and drawing out the elements necessary if we are to clear a pathway towards our destination. It also identifies some of the principal obstacles to taking this pathway. The following section examines how we can take this pathway. Returning to the issues of how scenarios could incorporate political economy approaches as outlined in Chap. 5, it focuses on how more robust political economy approaches can be fostered, looking at the role of the state, its relationship to market actors and the essential need for a creative and activist civil society. This is placed in the context of how political economy could contribute to exploring and creating the future in scenarios, and how this relates to seeking transition and transformation. The

P. Kirby, T. O'Mahony, *The Political Economy of the Low-Carbon Transition*, International Political Economy Series,
DOI 10.1007/978-3-319-62554-6_10

compelling and inseparable guardrails of social justice and ethics are then discussed before moving to the potential use of political economy models for transition, and some of the policy approaches that could be applied.

Yet, even if countries were to develop political economy models that set out a development pathway to the destination required, setbacks would happen and obstacles would be encountered. Section three examines what could be done to keep to the pathways laid out in the face of such difficulties. It thus assesses the prospects for following low-carbon pathways. It examines the conditions being created by the emerging global regime being put in place through the UNFCCC process, and given legal weight by the Paris Agreement, which requires reassessment each five years of progress to a low-carbon future. It assesses how surprises can arise to sideline this process, such as the election of the Trump administration in the US in late 2016, and what can be done to strengthen the resilience of the process in this situation. Finally, it reminds us of the futures awaiting humanity if we fail to create adequate pathways to a low-carbon future, and to keep to them.

Options: Clearing Pathways

In this book we have deliberately shifted the discussion from mitigation and adaptation, which have tended to structure responses to climate change, to place the focus on pathways. There are two main reasons for this: firstly, recognising that the principal task is to construct pathway(s) to a post-carbon society. This recognition frames consideration of the range of approaches required for mitigation, and to aid adaptation, including issues often ignored, such as fundamental changes in consumption practices in industrialised countries. Secondly, in keeping with the attempt in Chap. 1 to identify more precisely the many dimensions of the 'wicked problem' we face, mapping out pathways allows us keep in mind that climate change is only one of the planetary boundaries we are crossing. Ultimately, the challenge humanity faces is to find forms of social living that can flourish comfortably within these planetary limits; therefore the objective of keeping global warming to less than 2°C could distract from decisive action on other urgent dimensions such as biodiversity, land use and biogeochemical flows.

A focus on pathways prompted us to identify that one of the most serious problems to be addressed is what in Chap. 1 we called the 'disjuncture between the scale of the crisis and the poverty of responses to it'. Already in that chapter we identified the dominance of a technological paradigm as one of the principal constraints to addressing adequately the multidimensional nature of the cluster of problems we tend to group under the heading of 'climate change'. Chapter 2 identified more fully, in the dominant ways in which these problems are analysed and measured, the limits and hidden assumptions of the dominant paradigm, what we described as 'a strong bias towards technocratic solutions and inadequate consideration of economic, political and social dimensions of the transitions required'. Arguing that any pathways of social change begin within historical trajectories and structured socioeconomic and political systems, the chapter showed how paradigms are framed by the interrelationships of political and economic power; in other words they exist within political economy configurations. Finding pathways adequate to the challenges we face, therefore, requires moving beyond the confines of a dominant technological framing of the problems and our responses, and bringing in the wider political economy context that shapes them. In examining the social dimensions of the challenges, Chap. 3 introduced the concept of development pathways, namely the social, cultural and institutional factors that determine what type of development is pursued and the values informing it, including the type of economy, society, technology and environment this entails and the role for governance. Yet, it argued that the human factors involved are insufficiently explored in the transition literature, which frequently resorts to generalisations about social, cultural and institutional factors without substantive discussion. It showed how socio-technical transitions, a prominent approach to understanding transition, require being broadened to include consideration of wider social change.

This then opened the way for a discussion of development pathways in Sections II and III. Section II draws lessons from a survey of international development experiences in Chap. 4, and from how scenario studies treat the issues involved in Chap. 5. Section III then empirically examines the actual pathways being taken by developed countries and some key emerging economies in Chap. 6 and by developing countries in Chap. 7. By drawing on international development theory and practice to help clarify developmental choices as we attempt to transition to a low-carbon society, Chap. 4 introduces the concept of development models, how they are constituted and their social outcomes. It is shown how development pathways coalesce around models constituted by particular state-market-civil society relationships and this conceptual framework is applied to the challenges of climate change. The chapter also critically examined the concept of 'sustainable development' concluding that there is little evidence to show it has spurred decision-makers to consider the radical changes required in social, economic and political organisation if development is to be made truly sustainable. Chap. 5 turns to the use of scenarios for insight into the future, and how they have come to dominate our approaches to the low-carbon transition. The different techniques used in scenario studies are reviewed and the chapter than examines what environmental and emissions scenarios tell us. In critically assessing transition and transformation it is argued that a more fundamental transformation is needed beyond the limited reliance on techno-economic measures, to include sustainability and social, cultural and political drivers. This would not only facilitate a faster and more complete reduction in emissions, but also the achievement of potential development win-wins.

The extent to which the actual development pathways followed by countries around the world have managed to combine development with sustainability is the subject of Chaps. 6 and 7. While the steady if slow process of global climate diplomacy has placed global warming and its impacts on the political agenda of countries worldwide, they are addressing these challenges in different ways. Chap. 6 looks at two groups that have prioritised conventional economic development: the 'developed world' of the EU, the US and Japan, and the 'emerging economies' of China, India, Brazil and Mexico. A brief survey of the Nordic countries also highlights some of their successes. Focusing on identifying the principal means through which countries are addressing the related challenges of future transition and sustainable development, the chapter assesses how well they are doing and finds the techno-economic approaches adopted, while necessary, are insufficient and there is a need for sustainable development pathways and integrated development policy to enable successful transition. Chap. 7 focuses on developing countries and how they seek to integrate sustainability within their development pathways. It firstly draws on a survey of which developing countries manage to combine development with sustainability and highlights the success cases of Costa Rica and Uruguay. It goes on to highlight how, in all developing regions, their development gains are highly vulnerable to the impacts of climate change and identifies some of the particular responses from each major region: environmental activism in Latin America, the 'green state' in Africa which offers examples of combining development and sustainability, the pioneering of 'green growth' by some Asian countries, and the successes of diplomatic efforts by small island developing states (SIDS). The conclusions examine how state capacity is built up and the importance of political commitment, highlighting the role of civil society activism in shaping and reshaping states. The overall conclusion drawn from the two chapters is that the political economy models fashioning pathways towards a post-carbon society in most parts of the world display far too much deference towards market actors, devote inadequate attention to the development of state capacity and leadership commensurate with the immense challenges involved, and fail to develop the sorts of state-civil society partnerships that might galvanise the radical shifts needed.

Section IV turns to development models for a post-carbon society. Chap. 8 identifies the dominant model currently in place to achieve this transition, namely a climate capitalism while Chap. 9 outlines key dimensions of an alternative model that is emerging, which it labels ecosocialism. Examination of climate capitalism begins with a definition and a description before outlining examples in the reports of the New Climate Economy project, and policies based on green growth. The chapter then examines in turn three major challenges for climate capitalism: the key question of finance; how to govern the market so that it delivers the emissions' reductions necessary; and the fundamental challenge of decoupling growth from GHG emissions, a necessary condition for its success. The final section raises questions about the central role played by the free market and about economic growth itself. In looking for examples of alternatives to climate capitalism, Chap. 9 identifies emerging strategies of transformation in different sectors of the economy and society, assessing how these might be more successful in fashioning pathways towards a post-carbon society. The chapter goes on to outline debates on the need to move to a degrowth economy and society, what this might entail and how feasible it is. It then turns to one of the great problems of contemporary capitalism namely the structural nature of socioeconomic inequality and how to reduce it. It maps out an alternative 'economy for the common good', based on different social values and governance practices. The chapter therefore describes elements of an emerging alternative political economy model and finds in the utopian socialists of the early nineteenth century many of the principles motivating it. It is therefore being called 'ecosocialism' by some analysts. The chapter ends by assessing the benefits of climate capitalism and ecosocialism in moving us towards a post-carbon future, and a scorecard for each is drawn up.

So what pathways has the discussion so far helped to clear? The first is the need to move beyond the confines of a dominant technological framing both of the problems to be addressed and of the means to address them. We have identified this as a major obstacle to making decisive progress towards a post-carbon society because it fails to devote sufficient and rigorous attention to the wider developmental challenges involved. We have argued the need for development pathways that would redress the predominance of technological means in current scenario studies, requiring more comprehensive engagement with the social, cultural and institutional complexities of development. Examining the lessons that have been learnt from international development, we have highlighted the political economy structuring that underpins all development pathways, focusing attention on the need to examine the ways in which state, market and civil society combine and interrelate to achieve development. Our survey of the actual pathways being taken by countries towards a low-carbon future identified the role of the state as being, in most cases, insufficient to ensure that economic and social development are kept to a pathway towards a low-carbon future. As a result, the radical steps needed to advance on such a pathway are not being taken while the interests of capital accumulation appear in most cases to obscure the pathways that are needed. Civil society is incubating numerous examples of how to advance on these pathways but these seem in many cases to lack the support of more powerful economic and political interests. Yet, they elucidate a set of clear pathways that need to be recognised more widely as the only hope of achieving a low-carbon and sustainable society by 2050 and decisive steps taken to follow them. What are the prospects that this might happen?

Means: Taking These Pathways

Establishing the means of taking pathways towards a post-carbon and sustainable society draws on many of the lessons of the previous chapters in understanding and initiating the transition known to be required from the science. Chap. 4 discussed lessons learned from decades of international development. Chap. 5 discussed how the approach to developing scenarios of the future has led to greater realisation of the importance of social, cultural and political drivers in determining these outcomes, how political economy directly relates to these and how development pathways can articulate holistic visions of what they entail rather than just technological transitions. In Chaps. 6 and 7 we sought to understand the actual development pathways being pursued in different countries as outcomes of the development process. Chaps. 8 and 9 framed these outcomes more specifically as political economy models. But as we look to the future, in both analysis and policy, what is the place for political economy?

Scenarios

As the use of scenarios within the realm of climate change and transition has expanded, a greater inclusion of the social sciences in general, and political economy in particular, would not only represent a step forward in the practice of scenario development, but a potential renewal of political economy in the face of what many, such as former UN Secretary-General Ban Ki Moon, have described as the greatest threat that humanity has ever faced.[1] The review of energy scenario studies by Soderholm et al. (2011) argued that the continued reliance on quantitative energy scenarios by computer models has led to a rudimentary approach to politics, institutions and governance, a salient weakness, but that there is no fundamental obstruction to including these issues through qualitative scenario exercises that are complimentary and explore the societal transition. While scenarios can lean towards a cold technical analysis of development challenges such as transition, through illuminating discrete dimensions of the problems, they can also flag warning signs of what may lie around the next bend.

In the context of threats to sustainability, the Tellus Institute scenarios (Raskin et al. 2010) offer a useful exploration of alternative potential outcomes. The operative word here is 'alternative', offering hope that there are pathways open that lead to futures that could flourish with human and environmental wellbeing, as well as those that could lead to wrack and ruin. To respond to the need for research on practices for facilitating transformative change going beyond existing scenario approaches, Sharpe et al. (2016) developed the 'three horizons' approach that allows participants to work towards pathways that not only simplify the complexity of the subjects, but actively 'generate agency' and make explicit the power relationships this involves. There remain opportunities for societies to choose pathways that balance development, encouraging us not just to look at how things are, but at how they could be. Albeit, it must also be acknowledged, that the window of opportunity is narrowing and the passing of time will prove the enemy of further procrastination.

The case studies of individual countries in Chaps. 6 and 7 has allowed us to focus on actual development outcomes, and planned activities for low-carbon transition into the future. There are four common themes that we can distil from these case studies: (i) all countries need to increase the focus and rate of technological transition to a post-carbon future; (ii) this technological transition is not enough, all countries need to seek sustainable development pathways that balance social, environmental and economic outcomes; (iii) there are many win-wins and opportunities available through integrated policy approaches; (iv) the policy approach is crucial in determining what kind of outcomes are ultimately delivered. This fourth point is key for political economy. The IPCC has described the importance of 'governance' as a combination of state, market and civil society essential to define sustainable development paths (Sathaye et al. 2007: 697) and recognised social, political and cultural factors such as poverty and social equity as dimensions that cannot be separated and ignored. This dovetails with political economy as such relationships and outcomes are at the heart of both the challenges and the opportunities. A political economy approach would therefore be useful in the scenario development process as part of interdisciplinary teams that consider the evolution and interaction of driving forces, as advocates of an increased focus on social, political and institutional driving forces, in discussing different state-market-civil society patterns that may emerge, in seeking to understand where greater agency can be generated, in the assessment of the social and political implications of future scenarios and in the political discussion of the feasibility and implementation of policies and strategies towards transition. There is also a strong argument to be made for the use of scenario approaches within political economy, not only to aid consideration of transition, but to examine core questions in the field around the potential for change in the political and economic organisation of the future.

Transition to Transformation

In moving away from what the IPCC terms 'skewed development' (IPCC 2012: 37) or from what Nicholas Stern described as 'the greatest example of market failure we have ever seen' (Stern et al. 2006: 1), the scope of transition and transformation, and of the political economy of the necessary changes, comes into view. Transition tends to be more an incremental process of efficiency in current systems,[2] whereas transformation tends to be related to more fundamental change. However, transformation is not always desirable in and of itself, and ethically it is necessary to consider the means and the outcomes of what is envisaged. Yet, while the IPCC cautions that transformation can create a sense of disequilibrium and uncertainty, it also notes that transformations are already occurring at 'an unprecedented rate and scale, through globalisation, social and technological development, and environmental change. ... Climate change itself represents a system-scale transformation that will have widespread consequences for ecology and society, including through changes in climate extremes' (IPCC 2012: 466). Therefore the approaches should not be neatly defined as either transition or transformation but as both. They must be appropriate to the issue at hand and seek a political balance of the rights, interests and values of different stakeholders.[3] They should seek to direct development on a desired path but also respond to new developments as they occur. To meet ethical requirements, they must balance rights, interests and values, but they also must be democratically informed through public participation and political deliberation at multiple scales from local, through national to international.

Transformational responses can be facilitated through learning processes, especially reflexive learning that explores blind spots in current thinking. This is an opportunity to refresh our visions of what kind of societies we wish to achieve and to re-interpret the values this involves. It creates a need for state, market and civil society to establish a collective vision of a sustainable future, and crucially to empower the voices of the marginalised, future generations and the environment (see Box 10.1 on a Swiss initiative). It is acknowledged that vested interests seldom choose transformation, particularly when there is much to lose from change (Christensen 1997). As noted by Newell and Paterson (2010), the prevailing organisation of the global economy confers significant power on actors associated with fossil fuel interests and with the financial sector. There are winners and losers not only from extreme climate and weather events, but also from responses to them. These are inherently political questions, in which the power of the fossil-fuel lobbies has often been a dominant voice. Helping people, groups, organisations and governments to manage the resulting disequilibria is seen as essential to successful transformation (IPCC 2012: 466). Newell and Mulvaney (2013) referred to these issues as the political economy of the 'just transition', that addresses the need for climate justice across countries and generations, with the need for justice in access to energy for those currently in poverty while also addressing job losses in carbon-intensive industries.[4]

Box 10.1: Swiss Begin Discussion on Living Within Planetary Limits
In September 2016, Switzerland became the first country to vote on whether to implement a green economy. The initiative encouraged resource efficiency and moving to a circular economy. Specifically it set the goal of reducing Switzerland's resource consumption to a level that would require no more than one Earth by 2050; currently Swiss consumption requires the resources of 2.8 Earths.

An initiative of the Green Party, the proposal was opposed by a large majority in parliament and by the country's government. The environment minister, Doris Leuthard argued that 'the initiative calls for too much in too short a time'. Despite this, on a turnout of 42.5%, 36.4% voted Yes and only the canton of Geneva had a majority in favour. Despite the loss, Green Party president Regula Rytz said that the green economy was the winner because 'an important discussion about the future has begun' (Bechtel 2016).

Social Justice

The challenge of social inequality is a key plank of sustainable development pathways. It relates not just to social justice, human wellbeing and systems of political and economic organisation, but to the impacts of human activities on the environment.[5] Acknowledging that there are unequal outcomes in development, which can be remedied by policy, reflects the importance of the social dimension in considering transition and a sustainable society as interrelated goals. The OECD report *In it together: why less inequality benefits all* (OECD 2015) highlights that as inequality is continuing to grow in most OECD countries economic growth is damaged. More importantly, this not only deepens problems for those in poverty, it is damaging to society as a whole (Bartolini 2014). While global inequality generically is perceived to have declined, inequality between rich and poor countries and between the rich and poor in most countries has actually increased (Grimalda et al. 2016). For political and economic organisation, the lessons of history from the 'Great Depression' of 1929 to 1939 are pertinent here (Box 10.2).

Roosevelt's 'New Deal' is credited with either causing or accelerating recovery in the US, as it escaped the recession and addressed damning levels of inequality, unemployment and poverty. Many of the provisions to

Box 10.2: The 'Great Depression' and Roosevelt's 'New Deal': A Useful Echo of History on Political and Economic Organisation?

Beginning with the stockmarket crash in New York in 1929, the human cost of the 'Great Depression' was significant, driving poverty and deprivation and spreading globally. Many schools of thought explained the Great Depression, from Keynes on declining expenditures and unemployment, to the monetarists Friedman and Schwarz, who attributed it to the banking crisis and monetary contraction, through Irving Fisher's debt and deflation and Marxist boom-and-bust capitalism which generated inequality of wealth accumulation.

Two economists, Catching and Foster, proposed that inequality drove production at the expense of income and consumption. They recommended redistribution of wealth and public works, and their recommendations were followed by Presidents Hoover and Roosevelt. Regardless of the attribution of the exact causes, the free-market neoliberal style view of a self-correcting economic system proved false, and a major political realignment occurred. Roosevelt's 'New Deal' focussed on the '3Rs' of social security programmes that provided 'relief' for those in poverty, 'recovery' of the economy and 'reform' of the financial system. It involved protecting the organisation of labour, setting minimum wages, establishing public works, protecting migrants and ethnic minorities and separating commercial and investment banking to prevent speculation. This was effectively a model of economic and political organisation which sought to recognise what Grimalda et al. (2016) have described as both the positives and negatives of capitalism, to wed it to a progressive 'social ethos'.

address the Great Depression and prevent future crises were dismantled during the Reagan and Thatcher era of the 1980s, with waves of 'free-market' deregulation, lower taxes and cutbacks in the welfare state and public services in other countries following in their wake. However, levels of inequality in the US in 1928 are now matched today. With the 'Great Recession' of 2007–2008, the banking and financial crisis and seismic political tremors in the developed countries, similar problems are evident once more. A notable difference is that economic growth at all costs is no longer a viable option. It is not conducive to delivering the kind of progressive future of human and environmental wellbeing that the global majority wishes to achieve. In a world where the realities of climate change have begun to bite, a fossil-fuelled 'brown-growth' that occurred globally throughout the twentieth century, in moving from depression to affluence, will not deliver a sustainable future. A sustainable development pathway that addresses both the social and the environmental dimensions, and that transitions to a low-carbon and socially sustainable future, is now patently necessary in the twenty-first century.

In considering the example of social and economic inequalities, we begin to understand the place of ethics in development. The IPCC (Fleurbaey et al. 2014: 297–298) discuss these as issues of governance and political economy through:

1. rethinking the ways society relates to nature and the underlying biophysical systems;
2. complex intergenerational considerations;
3. fundamental restructuring of the global economic and social systems and,
4. sustainable development governance that cuts across several realms of policy and organisation.

These political economy challenges are linked to considerations of human and environmental rights and the right to development and human wellbeing. The ethics of environmental protection include 'ecocentric' views in strong sustainability, valuing the environment and nature in and of itself, and 'anthropocentric' views that tend to value only the 'ecosystem services' that the environment and nature provide to humanity. Issues of social justice arise both within countries and across countries and generations. The concept of *climate justice* emerged as an attempt to frame climate change as an ethical and political issue, noting that the worst impacts of climate change tend to be placed on those least responsible for the fossil-fuelled development that led to climate change occurring. The Mary Robinson Foundation—Climate Justice (MRFCJ), established by former President of Ireland and former UN High Commissioner for Human Rights, Mary Robinson, has been effective in linking climate change to human rights and development, to achieve a human-centred approach. It seeks to safeguard the rights of the most vulnerable and share the burdens and benefits of climate change and its impacts equitably and fairly. The MRFCJ has lobbied at UNFCCC meetings to place greater emphasis on this human-centred approach in global climate diplomacy, citing gender, poverty, inequality and the reality of the interdependence of all global citizens in addressing the climate issue.

Political Economy

Political economy models can function as a framework to implement the kind of transition and transformation that is required. The potential of the technological transition has been reviewed in Chap. 5, and its application in political economy through the 'climate capitalism' of Newell and Paterson (2010) reviewed in Chap. 8. Climate capitalism has limitations in securing a sustainable development pathway. It relies on technology deployment and economic instruments such as carbon taxes and emissions trading. While emissions reductions have occurred they have not been at the speed or scale required. Chap. 8 points to difficulties in reconciling the imperative of capital growth with that of decoupling and reducing emissions. The future of capitalism is indeed not only under pressure due to the 'skewed development' that has led to climate change (IPCC 2012: 37) but because of threats to neoliberalism and capitalism from technological change (Mason 2015: 120, 123) and the eroding of the social contract, as capitalism in its current form is failing to deliver the just and sustainable society that people seek (Deakin et al. 2016). Acknowledging the place of vested interests in preventing the realisation of a post-carbon future requires a new development model to be implemented. In this model, the state would have greater power to direct market forces, and civil society would have a much more active role to play to ensure socially just and sustainable outcomes.

In looking at the political economy models identified in Chap. 4 in 'Eco-innovation', 'Environmental State' and 'Climate Capitalism', and adding to these Audet's models of transition that included 'managerialist technocentrist' and 'radical ecocentrist', (Audet 2014: 47), we can map out various approaches to addressing the challenge of transition. Figure 10.1 illustrates these models differentiated by the axes of regulatory and policy intervention versus free market, and top-down versus bottom-up innovation processes. Sustainable development is compatible with any of these models if the outcomes are indeed socially, environmentally and economically sustainable.[6] This is one of the core criticisms of sustainable development, as this flexibility is a conceptual strength but it can be a political weakness. Sustainable development is informed by evidence, but

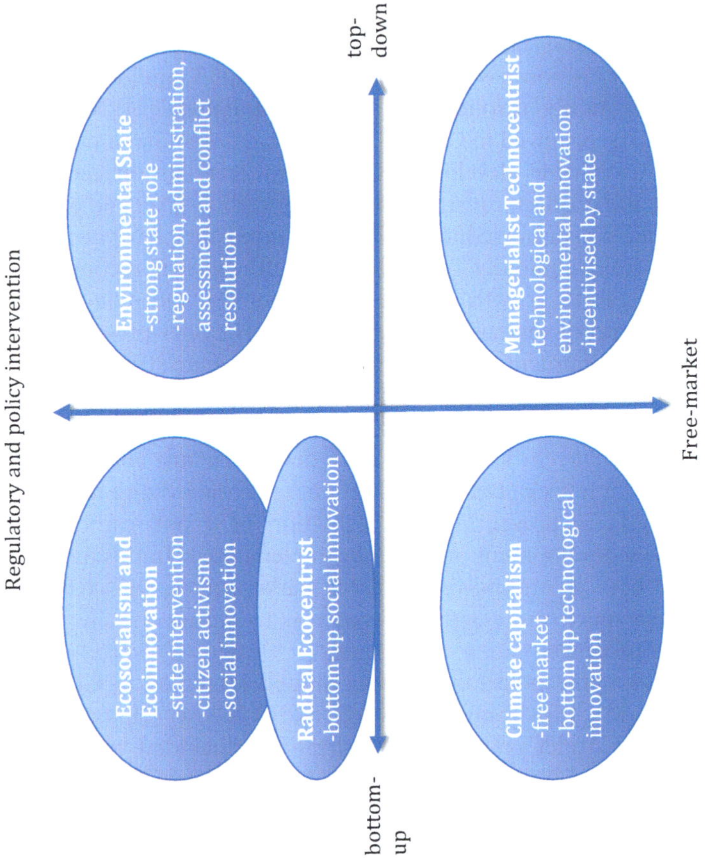

Fig. 10.1 An illustration of different political economy and transition approaches

it is not an absolute, it must be defined through political debate in each circumstance. However, normatively and in terms of its political approach, sustainable development is more compatible with state intervention and bottom-up social innovation, as this has a stronger correlation with guiding market forces towards positive social and environmental ends, and also in ensuring that there is equity and broad participation in balanced development.[7] While command-and-control strategies fell out of favour in recent decades and were overtaken by market-based instruments (Sathaye et al. 2007: 710), the World Bank has predicted that state intervention will inevitably play a greater role in directing development and transition (World Bank 2010: 61). An unswerving faith in the market's ability to deliver socially desired outcomes has been greatly shaken since 2008, similar to that which occurred during the Great Depression of the 1930s.

Framing political economy models for transition to a post-carbon future and a sustainable society can be aided by considering the Tellus Institute scenarios discussed in Chap. 5 (Raskin et al. 2010). While we are currently moving towards a 'Fortress World' scenario of an authoritarian path in response to mounting crises and insecurity, we have the potential to move to a 'Policy Reform' path of government-led redirection of growth towards sustainability goals, or indeed to a 'Great Transition' pathway as a fundamental transformation. This would involve a regulatory environment that incentivises market actors to be innovative and play their part, an activist civil society that both models best practice and puts pressure on government for the policies and leadership needed from the political system, and visionary political leadership. Such political leadership must deliver on all of our sustainability objectives and values and not just economic growth[8] and technological change. Policies that address the development pathway, and seek to achieve win-wins, acknowledge that the ends of development should be human and environmental wellbeing, with the economy and technology only as means. Ultimately delivering a sustainability and low-carbon transition depends on good governance such as 'whole-of-government decision-making, synergies among economic, environment and social policies, coalition building, political leadership, integrated approaches and policy coherence' (Sathaye et al. 2007: 717). But as stated by the IPCC, public policies alone cannot trigger changes in pathways, and cooperation between governments, markets, and civil societies is necessary. This requires participation in decision-making (Fisher et al. 2007: 178), and an active role for civil society.

Civil society can play a vital role in shaping development, encouraging empowerment and participation in decision-making, creating and influencing policy and engendering systemic grassroots change (Moser and Dilling 2007). These are the functions of civil society as social movement, as political movement and as a source of social innovation that models and experiments with sustainable society. In terms of the market, the role of industry in sustainability is clear and can be incentivised to encourage efforts to move towards industry and business sustainability. Drivers of corporate environmental sustainability include the capacity to influence regulation, green-marketing, managing stakeholder relations, the demands of investors, the demands of insurers and other financial institutions and stakeholder initiatives (Sathaye et al. 2007: 711–713). But the results of corporate sustainability and Corporate Social Responsibility (CSR) efforts could be described as decidedly mixed, and there are economic incentives to free-ride at best (in continuing to pollute, to encourage higher consumption and to work solely towards profit accumulation) and to prevent socially progressive action at worst (in the case of the fossil-fuel lobby seeking to misrepresent the science of climate change and weaken policy). It is necessary to have scrutiny and influence on market actors from state and civil society. This could prevent the role of industry from collapsing under its contradictions as described in the future of capitalism by the International Panel on Social Progress (Deakin et al. 2016).

In moving towards such an approach, ecosocialism is offered as an alternative in Chap. 9. It is interesting to note, that through its focus on outcomes for human wellbeing rather than growth in and of itself, ecosocialism is similar in its critique to that of mainstream development literature including the Nobel Economics Laureate Amartya Sen (Anand and Sen 2000).[9] In that sense, ecosocialism would not actually be a radical response, but a necessary discussion of the outcomes of development which are benefitting the few at the expense of the many, and also at the expense of the environment on which we depend. In Chap. 9, ecosocialism has been linked to the degrowth movement, to reducing materialism and consumerism, to the circular economy, moving to the 'economy for the common good' and to a socially cooperative rather than profit maximising market economy. Such an economy could respond to the concerns for the future of capitalism raised by the International Panel on Social Progress, and to the seismic political changes occurring in western countries since the recession of 2008, to which we will return in the next section. Jakob and Edenhofer (2015) offer an alternative view that critiques degrowth and green growth by suggesting that it is not a focus on growth that is important, but the use of a 'welfare' approach. This would address the negative aspects of the over-use of limited resources and over-consumption by the affluent, through shifting the focus to societal goals that address the decline of natural capital but also insufficient investments in the capability of people. They argue that the necessity of this dual approach is particularly in evidence with persistent extreme poverty in developing countries, requiring an increase in consumption and growth in these cases.

Policy Cycles

While elements of ecosocialism as a political model may be subject to debate (Burkett 2006: 23), similar to sustainable development pathways and the politics of problem-solving in general, there is scope to apply different aspects of ecosocialism in different contexts as part of discussions on what is nationally appropriate. Reflecting political and cultural differences, the IPCC alluded both to pragmatic political realities and the ethical requirements of establishing policy across diverse political and cultural contexts by stating:

> A substantial body of political theory identifies and explains national policy styles or political cultures. The underlying assumption is that individual countries tend to process problems in a specific manner, regardless of the distinctiveness or specific features of any specific problem; a national 'way of doing things'. (Sathaye et al. 2007: 709)

Recognising that sustainable development and low-carbon transition is a broad challenge that is not simply technological or economic, moves the focus towards sustainable development pathways discussed in Chap. 3. These can seek to deliver human wellbeing that enhances society and the environment. This involves an integrated vision for the development of societies that goes beyond technological transition, and recognises the importance of 'sustainable wellbeing' (O'Mahony 2016) and integrity of the environment on which we rely (see Box 10.3). In Chap. 3 we discussed the importance of carbon lock-in as development paths get locked in to particular directions. However more positively, Sathaye et al. (2007: 701) also describe a longer-term perspective in that the factors underlying the development path are subject to human intervention and are under the agency of governance. Our development patterns can be changed from economic growth and consumption to sustainable wellbeing, and to a sustainable society and environment that seeks to benefit all, and not just the few.

In policy terms it is now widely accepted that, while there may be trade-offs, there is also potential for beneficial synergies and win-wins. Policies may advance human wellbeing, reduce GHG emissions and enhance environmental integrity at the same time. Technology and energy policies can reduce the consumption of energy, and shift it towards renewables, with major benefits for public health through reducing air pollution. Policies that reduce over-consumption of meat can drastically reduce greenhouse gas emissions and reduce obesity, heart disease, diabetes and cancer, all scourges of public health in wealthier countries.[10] Policies for public and active transport, dense spatial planning, fostering communities[11] and environmental protection[12] can improve key inputs to quality of life. Such policies can also be used to re-direct economic development from a 'brown' to a 'green' model. This is balanced sustainable development, moving away from carbon- and materially-intensive products that seek ever higher turnover, regardless of the damage to the consumer or the environment. What may appear as more tangential policies also come into view as they have transformative potential. This could address the need to reconcile development with sustainability and move towards win-wins as previously discussed. Box 10.3 discusses one such approach in 'sustainable wellbeing'.

Box 10.3: A 'Sustainable Wellbeing' Approach to Improving Human Wellbeing and Reducing Material Consumption and GHG Emissions

Acknowledging that the high consumption of the wealthy is a barrier to transition (Fleurbaey et al. 2014: 308), has led researchers to consider what human wellbeing actually is and what other forms of wellbeing could be supported? A priority on income and consumerism can be damaging to our wellbeing, and it has been strongly argued that it is possible to have a more enjoyable and meaningful life through prioritising balance across life domains (O'Mahony 2016; Delle Fave et al. 2011).

The range of policies, measures and structures that could be used to facilitate this are expansive. Some could be identified as: universal income, shorter working weeks, and strong communities, families and personal relationships. In addition, addressing inequality and poverty, fostering appreciation of and access to the arts, music, education and nature offer alternative priorities. Supporting pro-social activities such as volunteering, the place of spirituality and religion, and priorities on physical and mental health, freedom and political empowerment are all dimensions that could be engaged. They could potentially initiate a step-change in human wellbeing, while reducing consumption and the environmental impacts of the affluent in a sustainable low-carbon society.

While, for those in poverty, incomes and consumption must increase (Jakob and Edenhofer 2015), balancing wellbeing amongst the more affluent is an opportunity that is largely ignored in discussions on transition. Breaking the perceived links between human wellbeing and consumption is a holy grail for improving sustainability. As a potential 'win-win', it is an enormous opportunity to facilitate greater human wellbeing at the same time (Jackson 2005).

Prospects: Keeping on the Pathways

The political events of 2016 with the election of President Trump, Brexit and the advance of the far-right in many Western countries seem defining features of the shifting political landscape of the early twenty-first century. The increase in inequality, the stagnation of living standards and wellbeing, and the increase in global insecurity arising from questionable foreign policies, have led to a scapegoating of immigrants, foreign cultures and those in poverty as the source of the slide. While the concerns of people in western countries are real and legitimate, and economic globalisation has been a double-edged sword, the retreat into nationalism and the political offerings evident in Trump's election and Brexit, where a core part of the narrative is to control immigration and scapegoat ethnic minorities, is a titanic failure to address the real causes of social and economic difficulties.[13] Countless reliable sources can be drawn on to show the economic and cultural contributions of immigration and even the lower crime rates among immigrants. The narrative that accompanies these developments seeks to further deregulate, lower taxes and weaken public services and the welfare state. It appears to be more about blame and fear, more ideological than rational, and it involves serious human costs. The wrong questions are accompanied by the wrong answers and the unfolding picture is not sustainable as social discord grows. Neoliberalism has to date effectively diverted attention from inequality, asymmetry of power and the growth of vested interests and the finance industry (including socialising banking debts and using public austerity to pay for it) from responsibility for social ills. Social and political movements and civil society activism become even more important as 'checks and balances' in situations where politics is not fully functioning in the interests of the people, but in the interests of capital and of the minority who are benefitting.

This turbulent period could lead to entrenching inequality as the real causes of social ills are obscured, or it could also lead to greater awareness and a determination to change course. This determination is strengthened as the major world religions seek to mobilise their followers to action (see Box 10.4). There is opportunity in these winds of

Box 10.4: World Religions Mobilise Action on Climate

The ground-breaking encyclical on care for our common home by Pope Francis is not the only statement by world religious leaders that seeks to mobilise their followers to address climate change and our socioecological crisis. Following the publication of the encyclical in June 2015, Islamic leaders meeting in Istanbul in August of the same year issued an Islamic Declaration on Climate calling on the world's 1.6 billion Muslims to play an active role in combatting climate change. They called for the phasing out of GHGs as soon as possible and a commitment to 100% renewable energy. 'Islam's teachings, which emphasise the duty of humans as stewards of the Earth and the teacher's role as an appointed guide to correct behaviour, provide guidance to take the right action on climate change', they state.

This was followed by the Buddhist Declaration on Climate Change in October 2015 calling for phasing out fossil fuels and challenging world leaders to close the emissions gap left by country climate pledges. It also welcomes statements from other religious traditions on the issue. The following month, a Hindu declaration called on the world's 900 million Hindus to make the transition to clean energy, adopt a plant-based diet, and lead lives in harmony with the natural world. Renewable energies provide the best option for billions of people to live better lives and reduce poverty, says the statement (McDonagh 2016).

change to move from the 'Fortress World' to the 'Great Transition' of Raskin et al. (2010). While it has been noted in scenario analysis that future change may be a challenge in the short-term, not only is it possible in the long-term, it is inevitable. This conclusion coincides with Grimalda et al. (2016) and their observation from history that democracy can transform capitalist institutions. In their view, the received wisdom today on capitalism as the end of history is a fallacy. It is both blind to the arc of history and the uncertainty and opportunity of the future. This uncertainty and opportunity is a staple of the scenario approach in considering possible future change.[14] The Trump administration, in choosing to ditch climate policy and attempting to reverse economic realities through a return to fossil fuels, is regressing not only from environmental sustainability but also from social and economic progress. In such an approach, the opportunities for future employment in clean industries will be lost to competitor countries and it is highly likely that US economic competitiveness will decline. If the 2016 US election victory suggests that the electorate sought a break from the inequality that emerged since the 1980s, then the tangent that the US is now evolving on could be described as tilting at windmills, a quixotic twist for a struggling majority in the world's richest country. While these political narratives continue they suggest a worsening of outcomes on all fronts without concerted opposition.

The global architecture and governance of the responses to climate change has also undergone much change. When the UNFCCC was established in 1992 it set in motion global political processes to reduce or 'mitigate' emissions. Even at this stage the science was sufficiently hard to warrant an urgency of action, and it was recognised that climate change was a collective problem which could not be resolved through the actions of individual countries. Urgency has only grown since as the impacts of climate change have escalated. There has also been a growing awareness of the potential for future damage or even breakdown in economic, social and environmental systems under the weight of climate change impacts. The Kyoto protocol mandated emissions cuts in developed countries up to 2008–2012. Despite the hardening of the scientific evidence, the recognition of current impacts and much discussion on the benefits of climate action, political developments in some countries ignored and conspired against climate action, pandering to a status quo that benefitted incumbent vested interests such as fossil-fuel industries. The Paris Agreement of 2015 sought to overcome this intransigence by compromise, establishing a legally binding treaty, but one that involves national flexibility in the setting of targets for reducing GHG emissions, and in the policies and measures that deliver on targets through INDCs. This could be described as politically pragmatic in the context of perceived difficulties in reducing emissions, and in moving away from fossil-fuelled development, deforestation and expansion of agriculture. However, it is unclear if it will deliver on the stated intention of the Agreement to limit the increase in global average temperature to less than 2°C and pursue efforts towards 1.5°C. No country could currently be described as being on a sustainable development pathway, although some countries are accelerating on the technological transition, and some such as India and some African countries have very low emissions per capita (albeit without strong outcomes in terms of human wellbeing, issues which must be addressed). The Paris Agreement includes a mechanism to increase the ambition of targets and plans[15] which responds to this requirement to accelerate the technological and sustainability transition, but whether it will be effective is an open question. Current transition plans are not compatible with reducing emissions to avoid dangerous climate change, and the INDCs submitted tend to focus only on the technological and not the sustainability transition.

International cooperation has had setbacks in the past as countries such as the US, Canada and Australia pulled out of Kyoto, often citing concerns over economic impacts on industrial competitiveness that don't stand up to scrutiny (Barker and Ekins 2004). Currently, the Trump administration is undoing progress, including ceasing the Climate Action Plan, and pulling out of the Paris Agreement or maybe even the UNFCCC. Such events could plausibly also have the opposite effect galvanising civil society in the US and international cooperation outside it. While the type of economic globalisation that has occurred in recent decades may have come with costs, globalisation of diplomacy through the UNFCCC and of science through the IPCC, has come with many benefits. The challenge of regressive political developments suggests that intergovernmental cooperation in the context of this collective action problem will need either robust mechanisms to deal with 'free-riders', or diplomatic efforts outside the UNFCCC, including potential trade restrictions.[16] The international dimension is patently necessary, as a chaotic national level approach will not drive global transition if some countries are intent on ignoring the unequivocal global scientific consensus on climate change. There is scope for discussion in what Fleurbaey et al. (2014: 298) allude to as collaborative learning so as to debate, legitimise and potentially overcome knowledge divides between experts and lay people. From a deliberative and procedural point-of-view, it is necessary to continue to dialogue and engage, and to address the misinformation and denial of science that has been evident. But it is also necessary to continue to act on the basis of the reality of the scientific evidence.

The alternative to effectively addressing climate change as a global phenomenon through international climate policy, and of failing to transition to a post-carbon world, are rapidly escalating climate change impacts (see Box 10.5). The cold term of 'climate impacts' hides a threat to our continued way of life, intensifying natural disasters with potential breakdown in social and economic systems. In order to understand the importance of transition, it is necessary to understand the future impacts of climate change. The IPCC has synthesised the outcomes of the science as climate change, risks and impacts (IPCC 2014). These synthesis reports are checked, tweaked and signed off line-by-line by the world's governments, and peer-reviewed by experts globally through an open process[17] while the conclusions are supported by the world's science academies. The conclusions do not equivocate on the gravity of the threat.

Box 10.5: Records Broken as Climate Reaches New Extremes

Even the period of writing this book (January 2016 to March 2017) saw evidence of growing extremes in climate changes and their impacts:

2016:

- *January*: Record cold temperatures and snowfall in large parts of East Asia. Twenty-four stations in China registered new lows while temperatures in Inner Mongolia reached a record low of −46.8°C. In Taiwan, at least 85 people were reported to have died of hypothermia.
- *February*: UN Food and Agriculture Organisation (FAO) announces that southern Africa is in the grip of an intense drought driven by one of the strongest El Niño events of the last 50 years. Large areas of Zimbabwe, Malawi, Zambia, South Africa, Mozambique, Botswana and Madagascar had the driest rainfall season in 35 years with 30 to 50 day delays in the onset of seasonal rains leading to widespread crop failure.

- *March*: A scientific study, published in the journal *Nature*, predicted sea levels could rise more rapidly than expected with collapsing Antarctic ice sheets expected to double sea-level rises by 2100.
- *May*: Temperature of 51°C recorded in Phalodi, India, on May 19th, the highest in India since records began.
- *May*: World Health Organisation reported that global urban air pollution levels increased by 8% between 2008 and 2013, a major cause of disease and death.
- *May–June*: A wildfire that began outside the city of Fort McMurray, Alberta, Canada, spread across 590,000 hectares (1,500,000 acres) before being brought under control in July. It destroyed 2,400 homes and buildings and a further 2,000 people were displaced after their homes were declared unfit for habitation. It was the most costly disaster in Canadian history and fires were still smouldering in early 2017.
- *June*: Intense rainfall in the UK, especially in London and the north of England, with repeated flooding in several locations on a daily basis.
- *July*: Temperature of 54°C recorded in Mitribah, Kuwait, on July 21st the highest on record for Asia. On July 22nd, Basra, Iraq, reached 53.9°C and Dehloran, Iran, reached 53°C, again breaking records.
- *August*: A massive flood swamped parts of southern and eastern Louisiana, the latest in a series of flood events that affected parts of Texas, Oklahoma, Louisiana, Arkansas and Missouri since March 2015. Tens of thousands of homes were flooded after 31 inches of rain fell in two days.
- *September*: Hurricane Matthew, the first Category 5 hurricane in the North Atlantic since Felix in 2007, hit Haiti, Cuba, the Bahamas and parts of southeastern US claiming more than 1000 fatalities and destroying thousands of buildings.
- *November*: Scientists express alarm at Arctic air temperatures peaking at 20°C higher than normal. While sea ice has declined by more than 30% over the past quarter century, it was at the lowest extent ever recorded for late November (Vidal 2016).

/...

2017:

- *January*: NASA and NOAA (National Oceanic and Atmospheric Administration) announce that 2016 was the warmest year in 137 years of record keeping and the third in a row to break that record.
- *January*: Chile faced what its President called 'the greatest forest disaster of its history' as fires destroyed more than 130,000 hectares (321,000 acres) and the country had to call in help from Brazil, Canada, Mexico, Peru and Spain to help fight them. It was followed by rainstorms and landslides that forced cutting off water to 4 million people in Santiago.
- *February*: Failure of late 2016 rains in parts of north and west Africa, coupled with civil strife, leads to UN announcing a famine in South Sudan. Some 70 million are expected to need emergency food aid in 2017 in an area stretching from South Sudan, across southern Ethiopia and northern Kenya into Somalia.
- *March*: World Meteorological Organisation (WMO) announces that 2016 was the hottest year on record pushing the world into 'truly uncharted territory' (Carrington 2017).

The risks include not only those to threatened ecosystems and animal and plant life, but to human health with increased disease vectors, water availability and drought, desertification, deforestation, threats to food production and security, coastal and urban floods, the inhabitability of low-lying and hotter locations, large-scale economic damages, threats to social organisation and increasing conflict as a 'threat multiplier'. Climate change also increases the risks of future natural catastrophes and disasters, and abrupt and irreversible changes to the natural systems on which we rely for survival. The potential magnitude of these impacts means that climate change is not 'another issue' to consider, but an existential threat which cannot be ignored. In a briefing to the UK parliament in 2010, Kevin Anderson and Alice Bows warned that we are on our way to 4°C warming 'an increase the UK's Committee on Climate Change considers to be extremely dangerous and incompatible with the contemporary framing of society and development' (Anderson and Bows 2010: 19). Anderson has even advocated for immediate and deep global decarbonisation rates of 10–20% per annum (from energy) to give a high probability of meeting a 2°C target, as opposed to most other analyses which advocate rates of typically 2, 3 or 4% per annum (Anderson 2012: 25–26). Transition is not just some interesting diversion or 'environmental cause' of a select few, it is a collective concern of humanity in general, with no rational or ethically justifiable opt-out.[18] Climate change is already occurring (see Box 10.5) and further change is inevitable, so that transition to a post-carbon future and adaptation to the impacts can't be avoided, and are patently sensible actions in the face of the weight of scientific evidence. To state in plain calculus, there will be economic costs in transition, but also many benefits which will significantly offset these costs. On the other hand, continuing on the current course will impose far higher costs (Stern et al. 2006) and risks to future generations and the environment that cannot be justified from an ethical standpoint.

Conclusion

The central problem facing humanity now is not that we don't know what we have to do to move our global society into balance with the carrying capacity of the planet on which it depends. This book has shown again and again that there exists a wide consensus on the direction we need to take. It was neatly summed up by two US biologists whose lifetimes' work makes them fear we are fast approaching irreversible and devastating tipping points. To avoid these, they say we need to do three things fast: reduce per capita consumption in the part of the world that is over-consuming (while allowing it increase for a period in poorer regions); getting real about the fact that 'the only thing that is going to work for the planet is maintaining the economy at a comfortable, consistent level, rather than constant growth'; and 'designing products whose cradle-to-grave environmental footprint is effectively zero' (Barnosky and Hadly 2016: 68). Not all would agree with this list and for low-carbon transition some would add the need to decarbonise the energy system, enhance carbon sinks, address deforestation and reduce other GHG emissions from industry and agriculture. However, what lists like these highlight is the central problem we face: that these measures seem today to be way beyond the capacity of our societies and their political systems to even begin in any decisive way.

Throughout this book, we have placed the focus on the limitations of the responses being made to the socioecological crisis we face, and sought to nourish a wider and more real discussion about how our highly complex and interconnected, yet deeply unequal and fragmented, global society needs to respond. We have placed these responses within the realities of economic and political power, outlining the contours of the sorts of development models we badly need to forge if we are to move towards where we must go. This chapter has mapped out in a comprehensive way the terrain of actions that are urgently required. Ultimately they take us back to the roots of politics: our individual and collective visions and the values that inform them, and how to create social structures that embody them. At heart the problem we face is a failure of political imagination and will, an inability to believe we are able to achieve the transformed society towards which our stressed planet is pointing us. As Mike Hulme has put it: ‘Climate change can help to bring the physical and the cultural, the material and spiritual, into a new realignment’, thus becoming ‘a mirror into which we can look and see exposed both our individual selves and our collective societies’ (Hulme 2009: 357).

A central feature of such a new realignment must be the ability to shake off the grip of a techno-economic paradigm that so limits our creativity and awareness, marginalising what is central to the constitution of all human societies: values, behaviour, consumption, power and politics, and active contestation. A realignment will therefore also lead to a reengagement, a new transformative political praxis that is visionary, ambitious and effective. Out of the contestation of ideas as reflected in this book must come a new politics or it is of little use, and the seeds of that politics can be dimly perceived in places (Simms 2016). While there is so much to be fearful about, ultimately the future will be shaped by the actions of individuals, communities, movements and parties able to articulate a credible vision and fashion a more humane, lively, creative and ecological future. Central to achieving this will be the ability to make technology and capital serve such a future rather than permitting the future be shaped by them to serve the needs of a small minority. Never before in human history has the future of humanity depended on the awakening and mobilisation of broad, progressive and generous social and political agency.

Notes

1. At the UN High-level summit on climate change in New York in 2014.
2. Incremental steps aim to improve efficiency within existing technological, governance and value systems, whereas transformation may involve alterations of fundamental attributes of those systems (IPCC 2012: 20).
3. Including that of future generations and the environment.
4. With potential for new employment in the low-carbon sectors.

5. Human impacts on the environment, such as the pressures of consumption of natural resources and the resulting increase in GHG emissions, are demonstrably higher for wealthier populations (Fleurbaey et al. 2014).
6. The concept of sustainable development is not unique in this respect. Its conceptual vagueness bears similarities to other norm-based meta-objectives such as 'democracy', 'freedom', and 'justice' (Sathaye et al. 2007: 697). It is argued that this allows different perspectives to engage in debate, and it effectively places the political process ahead of the political outcome.
7. Balanced development introduces an ethics but also a pragmatism that is necessary in an interconnected world. It would seek to address inequality in development, both within and between countries, to address poverty and improve human wellbeing. This is necessary for those experiencing poverty both in developed and in developing countries. It would also seek intergenerational justice that allows a similar level of opportunity and wellbeing in the future, and it would seek to protect the environment and nature for its services to humans and its intrinsic value in and of itself.
8. And the widening inequality and decline in public services that are currently occurring.
9. Jakob and Edenhofer (2015: 240) also agree with this point. In discussing the actual societal value of economic growth they state that 'economic growth cannot be a societal goal in itself, but can only be useful if it helps to achieve other objectives'. While on first reading this idea may seem politically revolutionary, it is a rather uncontroversial statement that forms the staple diet from the history of moral philosophy and much economic thinking and development literature.
10. This can also reduce the global competition and prices of staples such as cereals that are feed for livestock production, thereby improving the food security and nutrition of the global poor.
11. Cloughjordan Ecovillage in Ireland is an example of the integration of community, dense spatial planning with local services, public transport, renewable energy and economic development and employment, within a vision of local sustainable development and transition.
12. Biodiversity and habitat protection, expansion of native forestry, air, water and soil protection, national parks and conservation.
13. The narrative that has emerged also seeks to blame other marginalised groups including the unemployed and those on welfare benefits. Such political arguments are neither economically literate nor ethical and do not empower solutions.
14. Grimalda et al. (2016) propose that exploring feasible alternative systems may be one of the most important issues for the social sciences.
15. For the first time in the history of global climate policy, the Paris Agreement has established an ongoing, regular process to increase action by all

countries, known as the 'ratchet' or 'ambition mechanism'. From now until global temperature has been stabilised, countries must come together under the UNFCCC to review every five years, to take stock of progress and—informed by this stocktaking—submit a climate action plan that is progressively more ambitious.

16. For countries that repudiate even a highly flexible deal such as the Paris Agreement of 2015.
17. Suggestion that climate change science does not have a consensus is not supported. Science is a conservative process that functions by progressing debate through building evidence and seeking to come to conclusions. Once these core conclusions are sufficiently hard the debate moves on. This stage has long since passed for the science of climate change. Further research can refine some of the minutiae but the core conclusions, it can now be said unequivocally, will not change. Physics does not negotiate.
18. Even if the tolerance for risk may vary across people and nations, the magnitude of the risk and the potential for breakdown cannot be dismissed.

References

Anand, S., and A. Sen. 2000. Human Development and Economic Sustainability. *World Development* 28 (12): 2029–2049.

Anderson, K. 2012. Climate Change Going Beyond Dangerous—Brutal Numbers and Tenuous Hope. *Development Dialogue* 61: 16–40.

Anderson, K., and A. Bows. 2010. A 2°C Target? Get Real, Because 4°C is on its Way. *Parliamentary Brief* 13 (2): 19.

Audet, R. 2014. The Double Hermeneutic of Sustainability Transitions. *Environmental Innovation and Societal Transitions* 11: 46–49.

Barker, T., and P. Ekins. 2004. The Costs of Kyoto for the US Economy. *The Energy Journal* 25: 53–71.

Barnosky, Anthony D., and Elizabeth A. Hadly. 2016. *End Game: Tipping Point for Planet Earth?* London: William Collins.

Bartolini, S. 2014. Building Sustainability through Greater Happiness. *The Economic and Labour Relations Review* 25 (4): 587–602.

Bechtel, D. 2016. Green Initiative Will Not Leave Footprint on Economy. *Swissinfo*. Accessed 29 July 2017. https://www.swissinfo.ch/eng/september-25-vote_footprint-of-green-initiative-on-swiss-economy-/42465734

Burkett, Paul. 2006. Two Stages of Ecosocialism? Implications of Some Neglected Analyses of Ecological Conflict and Crisis. *International Journal of Political Economy* 35 (3): 23–45.

Carrington, Damian. 2017. Record-Breaking Climate Change Pushes World Into "Uncharted Territory". *The Guardian*, 21 March 2017.

Christensen, C.M. 1997. *The Innovator's Dilemma: When New Technologies Cause Great Firms to Fail.* Boston, MA: Harvard Business School Press.

Deakin, S., S. Stern, R. Kaplinsky, F. Muniesa, M. Nabli, M. O'Neill, H. Ortiz, K. Sahlin, A. Schwittay, and L. Talbot. 2016. Chapter 6: Markets, Finance and Corporations: Does Capitalism have a Future? Draft report for comment of the *International Panel on Social Progress.* Accessed 23 February 2017. https://comment.ipsp.org/chapter/chapter-6-markets-finance-and-corporations-does-capitalism-have-future

Delle Fave, A., I. Brdar, T. Freire, D. Vella-Brodrick, and M. Wissing. 2011. The Eudaimonic and Hedonic Components of Happiness: Qualitative and Quantitative Findings. *Social Indicators Research* 100 (2): 185–207.

Fisher, B.S., N. Nakicenovic, K. Alfsen, J. Corfee Morlot, F. de la Chesnaye, J.-Ch. Hourcade, K. Jiang, et al. 2007. Issues Related to Mitigation in the Long Term Context. In *Climate Change 2007: Mitigation Contribution of Working Group III to the Fourth Assessment Report of the Intergovernmental Panel on Climate Change*, ed. B. Metz, O.R. Davidson, P.R. Bosch, R. Dave, and L.A. Meyer. Cambridge: Cambridge University Press.

Fleurbaey, M., S. Kartha, S. Bolwig, Y.L. Chee, Y. Chen, E. Corbera, F. Lecocq, et al. 2014. Sustainable Development and Equity. In *Climate Change 2014: Mitigation of Climate Change. Contribution of Working Group III to the Fifth Assessment Report of the Intergovernmental Panel on Climate Change*, ed. O. Edenhofer, R. Pichs-Madruga, Y. Sokona, E. Farahani, S. Kadner, K. Seyboth, A. Adler, I. Baum, S. Brunner, P. Eickemeier, B. Kriemann, J. Savolainen, S. Schlömer, C. von Stechow, T. Zwickel, and J.C. Minx. Cambridge, UK and New York, NY: Cambridge University Press.

Grimalda, G., K. Moene, H. Ono, F. Filgueira, J. Roemer, R. Nanavaty, K. Gibson, C. Graham and D. Schkade. 2016. Chapter 8: Social Justice, Well-Being and Economic Organization. Draft report for comment of the *International Panel on Social Progress.* Accessed 23 February 2017. https://comment.ipsp.org/chapter/chapter-8-social-justice-well-being-and-economic-organization

Hulme, Mike. 2009. *Why we Disagree about Climate Change: Understanding Controversy, Inaction and Opportunity.* Cambridge: Cambridge University Press.

IPCC. 2012. *Managing the Risks of Extreme Events and Disasters to Advance Climate Change Adaptation.* A Special Report of Working Groups I and II of the Intergovernmental Panel on Climate Change [C.B. Field, V. Barros, T.F. Stocker, D. Qin, D.J. Dokken, K.L. Ebi, M.D. Mastrandrea, K.J. Mach, G.-K. Plattner, S.K. Allen, M. Tignor, and P.M. Midgley (eds)]. Cambridge: Cambridge University Press.

———. 2014. *Climate Change 2014: Synthesis Report. Contribution of Working Groups I, II and III to the Fifth Assessment Report of the Intergovernmental Panel on Climate Change* [Core Writing Team, R.K. Pachauri and L.A. Meyer (eds)]. Geneva: IPCC.

Jackson, T. 2005. *Motivating Sustainable Consumption. A Review of Evidence on Consumer Behaviour and Behavioural Change*. Guildford: Sustainable Development Research Network, University of Surrey.

Jakob, M., and O. Edenhofer. 2015. Welfare with or without Growth? Do We Need to Reduce Economic Activity to Protect the Environment and Increase the Quality of Life? *GAIA* 24 (4): 240–242.

Mason, P. 2015. *Postcapitalism: A Guide to Our Future*. London: Penguin.

McDonagh, Seán. 2016. Different Faiths United in Call for Climate Action. *The Irish Times*, 4 October 2016.

Moser, S., and L. Dilling. 2007. *Creating a Climate for Change: Communicating Climate Change and Facilitating Social Change*. Cambridge: Cambridge University Press.

Newell, P., and D. Mulvaney. 2013. The Political Economy of the 'Just Transition'. *The Geographical Journal* 179 (2): 132–140.

Newell, P., and M. Paterson. 2010. *Climate Capitalism: Global Warming and the Transformation of the Global Economy*. Cambridge: Cambridge University Press.

O'Mahony, T. 2016. Seeking Wellbeing in the Pathway of Low Carbon Transition. Plenary address to the *International Conference on Energy of Economic and Environmental Sustainability*, 21 October 2016, Lahore, Pakistan. doi:10.13140/RG.2.2.34339.25124

OECD. 2015. *In it Together: Why Less Inequality Benefits All*. Paris: Organisation for Economic Cooperation and Development.

Raskin, P., C. Electris, and R.A. Rosen. 2010. The Century Ahead: Searching for Sustainability. *Sustainability* 2: 2626–2651.

Sathaye, J., A. Najam, C. Cocklin, T. Heller, F. Lecocq, J. Llanes-Regueiro, J. Pan, et al. 2007. Sustainable Development and Mitigation. In *Climate Change 2007: Mitigation. Contribution of Working Group III to the Fourth Assessment Report of the Intergovernmental Panel on Climate Change*, ed. B. Metz, O.R. Davidson, P.R. Bosch, R. Dave, and L.A. Meyer. Cambridge, UK; New York, NY: Cambridge University Press.

Sharpe, B., A. Hodgson, G. Leicester, A. Lyon, and I. Fazey. 2016. Three Horizons: A Pathways Practice for Transformation. *Ecology and Society* 21 (2): 47. Accessed 24 February 2017. http://www.ecologyandsociety.org/vol21/iss2/art47/

Simms, Andrew. 2016. A New Type of Politics Could Help Prevent Climate Disaster. *The Guardian*, 9 September 2016.

Soderholm, P., R. Hildingsson, B. Johansson, J. Khan, and F. Wilhelmsson. 2011. Governing the Transition to Low-Carbon Futures: A Critical Survey of Energy Scenarios for 2050. *Futures* 43: 1105–1116.

Stern, N., S. Peters, V. Bakhshi, A. Bowen, C. Cameron, S. Catovsky, D. Crane, et al. 2006. *Stern Review: The Economics of Climate Change*. London: HM Treasury.

Vidal, John. 2016. "Extraordinarily Hot" Arctic Temperatures Alarm Scientists. *The Guardian*, 22 November 2016.

World Bank. 2010. *World Development Report 2010 Development and Climate Change*. Washington, DC: The World Bank.

Index

P. Kirby, T. O'Mahony, *The Political Economy of the Low-Carbon Transition*, International Political Economy Series,
DOI 10.1007/978-3-319-62554-6

D

The manufacturer's authorised representative in the EU is Springer Nature Customer Service Centre GmbH, Europaplatz 3, 69115 Heidelberg, Germany. If you have any concerns regarding our products, please contact ProductSafety@springernature.com

Printed and bound by CPI Group (UK) Ltd, Croydon, CR0 4YY
30/09/2026
02225652-0004